Alen-Pilip Prskalo

Molecular Dynamics Simulations of Si, SiC and SiN Layered Systems

Alen-Pilip Prskalo

Molecular Dynamics Simulations of Si, SiC and SiN Layered Systems

Südwestdeutscher Verlag für Hochschulschriften

Impressum / Imprint
Bibliografische Information der Deutschen Nationalbibliothek: Die Deutsche Nationalbibliothek verzeichnet diese Publikation in der Deutschen Nationalbibliografie; detaillierte bibliografische Daten sind im Internet über http://dnb.d-nb.de abrufbar.
Alle in diesem Buch genannten Marken und Produktnamen unterliegen warenzeichen-, marken- oder patentrechtlichem Schutz bzw. sind Warenzeichen oder eingetragene Warenzeichen der jeweiligen Inhaber. Die Wiedergabe von Marken, Produktnamen, Gebrauchsnamen, Handelsnamen, Warenbezeichnungen u.s.w. in diesem Werk berechtigt auch ohne besondere Kennzeichnung nicht zu der Annahme, dass solche Namen im Sinne der Warenzeichen- und Markenschutzgesetzgebung als frei zu betrachten wären und daher von jedermann benutzt werden dürften.

Bibliographic information published by the Deutsche Nationalbibliothek: The Deutsche Nationalbibliothek lists this publication in the Deutsche Nationalbibliografie; detailed bibliographic data are available in the Internet at http://dnb.d-nb.de.
Any brand names and product names mentioned in this book are subject to trademark, brand or patent protection and are trademarks or registered trademarks of their respective holders. The use of brand names, product names, common names, trade names, product descriptions etc. even without a particular marking in this works is in no way to be construed to mean that such names may be regarded as unrestricted in respect of trademark and brand protection legislation and could thus be used by anyone.

Coverbild / Cover image: www.ingimage.com

Verlag / Publisher:
Südwestdeutscher Verlag für Hochschulschriften
ist ein Imprint der / is a trademark of
OmniScriptum GmbH & Co. KG
Heinrich-Böcking-Str. 6-8, 66121 Saarbrücken, Deutschland / Germany
Email: info@svh-verlag.de

Herstellung: siehe letzte Seite /
Printed at: see last page
ISBN: 978-3-8381-3775-9

Zugl. / Approved by: Universität Stuttgart, Diss., 2013

Copyright © 2014 OmniScriptum GmbH & Co. KG
Alle Rechte vorbehalten. / All rights reserved. Saarbrücken 2014

List of Symbols

α_ν .... Intensity extinction coefficient [m^{-1}], page 124.

λ .... Wave lenght [m], page 123.

Å .... Ångstrøm, 1 Å = 10^{-10} m, page 14.

ω_{Larmor} .... Larmor angular frequency [s^{-1}], page 120.

$\vec{B}$ .... Magnetic field [T], page 120.

$\vec{E}$ .... Electric field [J/C];[N/m], page 120.

$\vec{F}$ .... Force [N], page 120.

a_{Bohr} .... Bohr radius a_{Bohr}=0.529 Å, page 36.

e .... Electron unit charge, $e = 1.602 \cdot 10^{-19}$ C, page 118.

eV .... Atomic energy unit, 1 eV = $1.602 \cdot 10^{-19}$ J, page 62.

h .... Plank constant 6.626 · 10^{-34} J·s, page 124.

I_d, I_0 .... Light intensity [J/m^{2}s], page 124.

n_0 .... Neutral atoms in a plasma, page 118.

n_i .... Ionized atoms in a plasma, page 118.

r_{Larmor} .... Larmor radius [m], page 120.

ADF	Angular Distribution Function, page 215.
AES	Auger Electron Spectroscopy, page 53.
AFM	Atomic Force Microscope, page 73.
AMOLED	Active Matrix Organic Light-Emitting Diode, page 141.
BCA	Binary Collision Approximation, page 64.
BOP	Bond Order Potential, page 14.
CSIRO	Commonwealth Scientific and Industrial Research Organisation, page 73.
CVD	Chemical Vapor Deposition, page 7.
DC sputtering	Direct current sputtering, page 117.
DD	Dislocation Dynamics, page 9.
DFG	Deutsche Forschungsgemeinschaft, German Research Foundation, page 7.
DFT	Density Functional Theory, page 9.
EAM	Embedded Atom Method, page 14.
EELS	Electron Energy Loss Spectroscopy, page 123.
EPMA	Electron Microprobe, page 122.
ESPResSO	Extensible Simulation Package for the Research on Soft Matter, page 26.
FEM	Finite Element Method, page 9.
FTIR	Fourier Transformation Infrared Spectoscropy, page 124.

GLOK .... Global convergence integrator, page 31.

HDMS .... Hexamethyldisilazane [(CH$_3$)$_3$Si]$_2$NH, page 54.

HF sputtering . High frequency sputtering, page 117.

HFFEM ...... Hartree Fock Finite Element Method, page 13.

HREM .... High-Resolution Electron Miscroscopy, page 141.

IAM-AWP .... Institut für Angewandte Materialien - Angewandte Werkstoffphysik, page 7.

IMWF .... Institut für Materialprüfung, Werkstoffkunde und Festigkeitslehre, page 7.

KIT .... Karlsruher Institut für Technologie, page 8.

LAMMPS ..... Large-scale Atomic/Molecular Massively Parallel Simulator, page 26.

LJ .... Lennard Jones potential, page 14.

MBE .... Molecular Meam Epitaxy, page 127.

MD .... Molecular Dynamics, page 7.

MEAM .... Modified Embedded Atom Method, page 34.

MIK .... Microconvergence integrator, page 31.

MMM .... Multiscale Materials Modeling, page 9.

NPT .... Great canonical ensemble, page 29.

NVE .... Microcanonical ensemble, page 31.

NVT .... Canonical ensemble, page 28.

PBC .... Periodic Boundary Conditions, page 32.

PDV	Physical Vapor Deposition, page 7.
PFM	Phase Field Method, page 9.
QMC	Quantum Monte Carlo, page 13.
RDF	Radial Distribution Function, page 76.
REM	Scanning Electron Microscopy, page 122.
RTA	Rapid Thermal Annealing, page 141.
SIMS	Secondary Ion Mass Spectroscopy, page 53.
TEM	Transmission Electron Microscopy, page 124.
TFT	Thin Film Transistor, page 141.
TRIM	TRansport of Ions in Matter - A Monte Carlo simulation package., page 64.
UMIS 2000 ...	Ultra Micro Indentation System. Nanoindentation system used in present work., page 73.
XPS	X-ray Photoelectron Spectroscopy, page 53.
XRD	X-ray Diffractometry, page 123.
XRR	X-Ray Reflectometry, page 122.
XTEM	Cross-sectional Transmission Electron Microscopy, page 76.
ZBL	Ziegler-Biersack-Littmark potential, page 35.

Table of Contents

	Page
List of Symbols	i
Table of Contents	v
1 Extended Abstract	1
1.1 Outline	2
2 Zusammenfassung	7
2.1 Aufbau der Arbeit	8
3 Introduction	13
3.1 Computational Materials Science	13
3.2 Description of the material system	18
3.2.1 Silicon carbide - SiC	20
3.2.2 Silicon nitride - Si_3N_4	21
4 Molecular Dynamics (MD) method	23
4.1 Quantum-mechanical approach	23
4.2 Differential equations	25
4.3 Algorithms for numerical integration	26
4.3.1 Verlet algorithm	26
4.3.2 Leapfrog algorithm	27
4.3.3 Velocity-Verlet algorithm	28
4.4 Thermodynamic ensembles	28
4.4.1 Temperature control	28
4.4.2 Pressure control	29

Table of Contents

		4.4.3	Micro convergence integrator	31
		4.4.4	Global convergence integrator	31
	4.5	Periodic boundary conditions		32
	4.6	Theory of interatomic potentials		33
		4.6.1	Pair potentials	34
		4.6.2	Metallic bond: Embedded Atom Method	38
		4.6.3	Covalent bond	39
		4.6.4	Bond ionicity	46

5 Simulation of the sputtering process — 53
- 5.1 Introduction . . . 53
- 5.2 Fundamental processes . . . 54
- 5.3 Sputtering mechanisms . . . 56
- 5.4 Sputter threshold energy and sputter yield . . . 58
- 5.5 Sigmunds sputter theory . . . 58
- 5.6 Simulation of the sputtering process . . . 64
- 5.7 Validation and extension of the sputtering results . . . 65

6 Simulation of the nanoindentation — 73
- 6.1 Introduction . . . 73
- 6.2 Nanoindentation of single crystals . . . 75
- 6.3 Nanoindentation of coating systems . . . 80

7 Results of MD sputtering simulations — 85
- 7.1 Computational details . . . 85
- 7.2 Sputtering process of silicon . . . 88
- 7.3 Sputtering process of silicon carbide . . . 96
- 7.4 Discussion and experimental validation . . . 108
- 7.5 Sputtering process of silicon nitride . . . 110
- 7.6 Discussion and validation by literature results . . . 114

8 Basic concepts of experimental PVD — 117
- 8.1 Experimental physical vapor deposition process . . . 117
  - 8.1.1 Plasma physics . . . 118

	8.1.2	Direct current DC sputtering	118
	8.1.3	High-frequency HF and magnetron sputtering . .	119
8.2	Experimental characterisation methods		121
	8.2.1	Measurement of the coating thickness	121
	8.2.2	Density analysis	122
	8.2.3	Chemical composition	122
	8.2.4	Microstructure and bond-type analysis	123
	8.2.5	Fourier Transformation Infrared Spectroscopy . .	124
	8.2.6	Raman spectroscopy	125
	8.2.7	Analysis of coatings mechanical properties	125

9 MD simulation of the deposition process   127

- 9.1 Description of the methodology 127
- 9.2 Deposition process of silicon on a Si substrate 129
  - 9.2.1 Silicon coating structure vs. deposition parameters 129
  - 9.2.2 Thermal annealing of silicon coatings 133
- 9.3 SiC deposition process on a Si substrate. 142
  - 9.3.1 Silicon carbide gradient coatings 149
- 9.4 Si_3N_4 deposition process on a Si substrate 151
  - 9.4.1 Silicon nitride gradient coatings 153
  - 9.4.2 Experimental observations 155

10 Nanoindentation of Si, β-SiC and α-Si_3N_4 single crystals   157

- 10.1 Application of the method by Oliver and Pharr 158
- 10.2 Nanoindentation of silicon 161
- 10.3 Nanoindentation of β-SiC. 175
- 10.4 Nanoindentation of α-Si_3N_4 183
- 10.5 Summary 189

11 Nanoindentation of SiC- and Si_3N_4-coatings   191

- 11.1 Summary 199

12 Summary and Conclusions   203

13 List of conference contributions and publications   207

Table of Contents

    13.1 Conference talks and poster presentations 207
    13.2 Journal publications 209

List of Figures   211

List of Tables   220

Bibliography   221

Acknowledgement   239

1 Extended Abstract

Experimental and empirical work in the field of nano-scale, functional multilayer coatings produced by physical (PVD) or chemical (CVD) vapor deposition are currently an object of international research. The research results have already been implemented in industrial products, however, the atomistic understanding of the underlying processes during deposition of these multilayer systems is still incomplete. In particular the correlation between the deposition parameters and the resulting coating structure needs further attention. To overcome the heuristic stage of the coating development and to facilitate targeted and efficient optimization of the coating structure and the resulting mechanical properties, a deeper understanding of the correlation between the substrate temperature, deposition rate and energy onto the growing coating structure is necessary. The same holds for the structure-properties correlation of these systems with a large number of interfaces as well as for the mechanical properties of individual layers.

The aim of this work is the combination and mutual validation of modern molecular dynamics simulations (MD) with experimental methods of physical vapor deposition of silicon and silicon-based SiC- and Si_3N_4-protective coatings. This was done within the framework of the research project *"Molecular dynamics modeling and validation of manufacturing and structure-property correlations of SiC/SiN nano laminates"* funded by the German Research Foundation (DFG) and in a collaboration between the Institute for Materials Testing, Materials Science and Strength of Materials (IMWF) at the University of Stuttgart and the Institute for Applied Materials - Applied Materials Physics (IAM-AWP) at the Karlsruhe Institute of Technology (KIT). While molecular dynamics simu-

Chapter 1. Extended Abstract

lations were performed at the IMWF, experimental investigations took place at the IAM-AWP. Hereby, the results of MD studies served as input parameters for experimental investigations. In this way, time-consuming and expensive experimental investigations could be avoided, since only optimistic simulation predictions had to be verified experimentally. In return, the group at the IAM-AWP delivered experimental results which were used to validate the simulation model.

In general, each experimental study was also represented by a MD simulation. Hence, the entire process of the coating deposition and characterization can be divided into three distinct stages:

- Simulation and experimental investigation of the sputtering process of Si, SiC and Si_3N_4.

- Simulation und experimental investigation of the deposition process of silicon substrates by Si, SiC and Si_3N_4.

- MD simulations and experimental characterisation of the correlation
between the structure and properties of the deposited silicon and silicon-based SiC- and Si_3N_4- protective coatings.

1.1 Outline

The present work is structured into 13 Chapters. **Chapter 3** provides an introduction into the Si-C-N material system, the existing crystal structures and mechanical properties of individual material subsystems. Simulation methods currently used in materials science are shortly described in the same chapter, e. g. Density Functional Theory (DFT), Molecular Dynamics (MD), Phase Field Method (PFM), Dislocation Dynamics (DD) and Finite Element Method (FEM). Finally, the concept of Multiscale Materials Modeling is also introduced in Chapter 3.

A more detailed description of the MD simulation method is given in **Chapter 4**. As a starting point, the Schrödinger equation is taken.

Following, the Born-Oppenheimer approximation, used for the separation of the mechanics of electrons and the nuclei is explained in the same chapter. The Tersoff potential, used throughout this work is also described in Chapter 4.

Chapter 5 provides a theoretical background of the sputtering process. Different sputtering mechanisms, such as the *single knock-on effect*, the *binary collision cascade* and the *thermal spike model* are explained and set into relation with the energies of the impacting ions. Also, terms such as the sputter threshold energy and sputter yield, both of fundamental importance for the understanding of the sputtering process are defined in Chapter 5. In addition, the sputtering model of Peter Sigmund, which is the most frequently referenced in the literature, is explained in the same chapter.

Experimental nanoindentation and its implementation in MD simulations are the topic of **Chapter 6**. Previous atomistic simulations and experimental results of the nanoindentation within the Si-C-N material system are cited and mutually validated. Comparison of both individual simulation results as well as the comparison with experimental investigations reveals substantial differences. These differences originate in different system sizes, different indenter forms and in the case of MD simulations, application of different interatomic potentials. Therefore, a systematic MD study of the nanoindentation of single crystals in the Si-C-N material system using the Tersoff potential was required.

Chapter 7 presents self-developed MD simulation models and experimental results of the sputtering of Si, SiC and Si_3N_4 by Ar^+ ions. Penetration depths of incident argon ions, the resulting forward and back sputter yield are analyzed and set into relation with ion impact energies and crystal orientations of the target material. The results of Molecular Dynamics simulations are validated by experimental measurements of argon-etched samples. Overall, an excellent agreement between experimental results and MD simulations could be achieved, in addition, available literature data could be confirmed.

Chapter 8 provides a summary over experimental methods currently

Chapter 1. Extended Abstract

used for the deposition of Si-, SiC- and Si_3N_4-layered systems. In addition, different characterisation methods used for the analysis of the chemical decomposition of deposited coating systems, the measurement of the thickness of individual layers as well as the analysis of their mechanical properties are presented.

The mutual validation between MD simulations and experimental investigations continued in the study of the deposition process is presented in **Chapter 9**. Doing so, it was possible to investigate the deposition process of Si, SiC and Si_3N_4 on a silicon substrate as a function of a large number of deposition parameters, such as the substrate temperature and crystal orientation, deposition rate and the deposition energy. In addition, a particle-dependent deposition rate was introduced and the influence of each parameter on the resulting coating structure could be determined.

In **Chapter 10** the mechanical properties of Si, SiC and Si_3N_4 single crystals were investigated by the method of nanoindentation using a Berkovich indenter tip. This was done systematically for the three low-index crystal orientations, namely (100), (110) and (111) in case of Si and β-SiC, and (0001), (10$\bar{1}$0) and (12$\bar{1}$0) in the case of α-Si_3N_4 single crystal. In this way, it was possible to evaluate nanohardnesses and Young's moduli of different material systems as well as to compare MD simulation results by available experimental investigations of the nanoindentation of Si, SiC and Si_3N_4.

Chapter 11 presents the investigation of mechanical properties of SiC- and Si_3N_4-coatings, deposited on a silicon substrate. The deposition process itself is described in **Chapter 9** while hardness measurements using nanoindentation are presented in **Chapter 10**. For the investigation of mechanical properties of layered systems, stoichiometric SiC- and Si_3N_4-coatings are used, while the investigation of mechanical properties of silicon as well as of SiC- and SiN-gradient coatings can serve as an outlook for further research. MD simulations of the nanoindentation are validated by the experimental nanoindentation using a UMIS 2000 nanoindenter, performed at the IAM-AWP at the KIT.

The results of this study, the measured hardnesses and the description of the substrate influence onto the measurement of the coating hardness can also be found in [18, 71].

A summary of the research results obtained in this work can be found in **Chapter 12**.

The results obtained in this study were presented at international conferences and resulted in three journal publications, the list can be found in **Chapter 13**. During the work at IMWF, contributions to the faculty teaching was carried out, as an example, a student research project (in German *Studienarbeit*) was supervised, while a Master Thesis is still in progress. Finally, the results presented in this work have also been incorporated into lectures at the University of Stuttgart (Prof. Dr. Siegfried Schmauder) and Karlsruhe (PD Dr. Sven Ulrich).

2 Zusammenfassung

Experimentell-empirische Arbeiten im Bereich nanoskaliger, funktioneller Multilagenschichten, die mittels physikalischer (PVD) oder chemischer (CVD) Gasphasenabscheidung hergestellt werden, sind derzeit internationaler Forschungsgegenstand. Die erzielten Forschungsergebnisse sind bereits vielfach in industrielle Produkte umgesetzt worden. Dennoch ist das atomistische Verständnis von Prozessen, die sich bei der Abscheidung solcher Schichtsystemen abspielen, weitgehend lückenhaft. Dies trifft insbesondere auf die Korrelation zwischen den Abscheideparametern und der resultierenden Schichtstruktur zu. Um das heuristische Stadium der Beschichtungsentwicklung zu verlassen und eine gezielte und effiziente Optimierung der Schichtstruktur zu ermöglichen, ist ein tieferes Verständnis der Korrelation zwischen der Substrattemperatur, der Abscheiderate und -energie und der daraus resultierenden Schichtstruktur mit hohem Grenzflächenanteil sowie den sich einstellenden mechanischen Eigenschaften erforderlich.

Ziel dieser Arbeit ist es, moderne Molekulardynamiksimulationen (MD) und experimentelle Methoden der physikalischen Gasphasenabscheidung von Silizium und siliziumbasierten SiC- und Si_3N_4-Schutzschichten miteinander zu kombinieren und gegenseitig zu validieren. Dies geschah im Rahmen eines von der Deutschen Forschungsgemeinschaft (DFG) geförderten Projektes *"Molekulardynamische Modellierung und Validierung der Herstellung und der Struktur-Eigenschafts-Korrelationen von SiC/SiN-Nanolaminaten"*, in einer Zusammenarbeit des Instituts für Materialprüfung, Werkstoffkunde und Festigkeitslehre (IMWF) der Universität Stuttgart mit dem Institut für Angewandte Materialien - Angewandte Werkstoffphysik (IAM-AWP) des Karlsruher Instituts für

Technologie (KIT). Die Zusammenarbeit der beiden Institute wurde so gestaltet, dass die Arbeitsgruppe am IMWF molekulardynamische Simulationen durchführte, während am IAM-AWP experimentelle Untersuchungen stattfanden. Die Ergebnisse der MD-Studien dienten als Eingabeparameter für die experimentellen Untersuchungen. Auf diese Weise konnten zeit- und kostenaufwendige Untersuchungen vermieden werden, da nur simulationstechnisch optimistische Voraussagen experimentell umgesetzt wurden. Im Gegenzug lieferte die Arbeitsgruppe am IAM-AWP experimentell relevante Ergebnisse, die zur Validierung des Simulationsmodells dienten.

Jede experimentelle Untersuchung wurde auch simulationstechnisch umgesetzt. Das gesamte Arbeitspaket lässt sich somit in drei Einzelpakete aufteilen:

- Simulation und experimentelle Umsetzung des Zerstäubungsprozesses von Si, SiC und Si_3N_4.

- Simulation und experimentelle Umsetzung des Beschichtungsprozesses von Siliziumsubstraten mit Si, SiC und Si_3N_4.

- Simulationstechnische und experimentelle Charakterisierung der Struktur-Eigenschaften-Korrelationen von abgeschiedenen Silizium- und silizium basierten SiC- und Si_3N_4-Schutzschichten. Eine besondere Bedeutung wurde der Analyse von Eigenspannungen und der Messung der Härte mittels Nanoindentation an den Schichtsystemen beigemessen.

2.1 Aufbau der Arbeit

Die vorliegende Arbeit ist aus insgesamt 13 Kapiteln aufgebaut. In **Kapitel 3** wird ein Überblick über das verwendete Si-C-N-Materialsystem, die vorliegenden Kristallstrukturen und die mechanischen Eigenschaften gegeben. Im selben Kapitel werden auch Simulationsmethoden, die derzeit in der Materialwissenschaft Verwendung finden; wie z.B.

die Dichtefunktionaltheorie (DFT), die Molekulardynamik (MD), die Phasenfeldmethode (PFM), die Versetzungsdynamik (engl. Dislocation Dynamics (DD)) und die Finite-Elemente-Methode (FEM), vorgestellt. Nicht zuletzt wird in Kapitel 3 das Konzept der Multiskalenmodellierung (engl. Multiscale Materials Modeling (MMM)), das derzeit zunehmend an Bedeutung gewinnt, eingeführt.

Die Molekulardynamik als Simulationsmethode wird in **Kapitel 4** im Detail beschrieben. Als Ausgangspunkt dient dabei die Schrödingergleichung. Desweiteren wird im selben Kapitel die Born-Oppenheimer-Näherung erklärt. Diese ermöglicht die Aufspaltung der Gesamtwellenfunktion in einen Elektronen- und einen Kernanteil. Ebenso wird in **Kapitel 4** das Tersoffpotenzial beschrieben, da es in dieser Arbeit durchgehend eingesetzt wird.

Theoretische Hintergründe des Zerstäubungsprozesses werden in **Kapitel 5** erarbeitet. Dabei werden verschiedene Zerstäubungsmechanismen, wie der *single-knock-on-Effekt*, die *binäre Stosskaskade* und das *thermal-spike-Modell* erklärt und in Verbindung mit den Energien der auftreffenden Ionen gesetzt. Auch werden Begriffe wie Zerstäubungsschwellenenergie und Zerstäubungsrate definiert, welche von elementarer Bedeutung für das Verständnis des Zerstäubungsprozesses sind und das erfolgreichste und meistzitierte Zerstäubungsmodell von Peter Sigmund erklärt.

Die Simulation und die experimentelle Umsetzung der Nanoindentation sind Gegenstand des **Kapitels 6**. Bisherige Ergebnisse der Simulation und der experimentellen Nanoindentation im Si-C-N-Materialsystem werden vorgestellt und miteinander verglichen. Dabei konnten Unterschiede, sowohl unter den einzelnen Simulationsergebnissen, als auch zwischen den Ergebnissen der numerischen und experimentellen Untersuchungen, nachgewiesen werden. Deren Ursprung lässt sich auf die Verwendung unterschiedlicher Systemgrößen und Indenterformen, sowie, bei den MD-Simulationen, auf unterschiedliche Wechselwirkungspotentiale zurückführen. Die Vielzahl unterschiedlicher Daten für die mechanischen Eigenschaften im Si-C-N-System motivierten zu einer

Chapter 2. Zusammenfassung

systematischer MD-Studie der Nanoindentation unter Verwendung des Tersoffpotenzials.

In **Kapitel 7** werden eigene Simulationsvoraussagen und experimentelle Ergebnisse der Zerstäubung von Si, SiC und Si_3N_4 mit Ar^+-Ionen vorgestellt. Im Einzelnen werden die Eindringtiefe der einfallenden Ar^+-Ionen, die Vorwärts- und die Rückwärtszerstäubungsrate, sowie der Anteil an zerstäubten Clustern analysiert. Die Ergebnisse der MD-Simulationen werden mit experimentellen Messungen argongeätzter Proben verglichen. Dabei konnte eine ausgezeichnete Übereinstimmung zwischen den experimentellen Ergebnissen und den MD-Simulationen erreicht werden. Zusätzlich konnten die partiell vorhandenen Literaturergebnisse bestätigt werden.

Kapitel 8 gibt einen Überblick über experimentelle Methoden, die bei der Abscheidung von Si-, SiC- und Si_3N_4-Schichtsystemen derzeit eingesetzt werden. Zusätzlich werden verschiedene Untersuchungsmethoden, die zur Analyse der chemischen Zusammensetzung abgeschiedener Schichtsysteme, zur Messung der Dicke von einzelnen Lagen oder zur Bestimmung der mechanischen Eigenschaften herangezogen werden, beschrieben.

Die gegenseitige Validierung der simulationstechnischen und experimentellen Untersuchungsmethoden wurde bei der Untersuchung des Abscheideprozesses fortgesetzt. So war es möglich, den Abscheideprozess von Si, SiC und Si_3N_4 auf einem Siliziumsubstrat, in Abhängigkeit von einer Vielzahl von Parametern, wie z. B. der Substrattemperatur, der Kristallorientierung, der Beschichtungsrate und der Beschichtungsenergie zu modellieren. Zusätzlich konnte eine partikelabhängige Beschichtungsrate eingeführt werden. Weiterhin konnte der Einfluss eines jeden Beschichtungsparameters auf die wachsende Schichtstruktur ermittelt werden. Die Ergebnisse dieser Untersuchungen lassen sich in **Kapitel 9** finden.

In **Kapitel 10** werden die mechanischen Eigenschaften von Si, SiC und Si_3N_4 mittels Nanoindentation untersucht. Dies wird systematisch für die drei wichtigsten Kristallorientierungen der jeweiligen Kristallstrukturen, nämlich für die (100), (110) und (111) im Falle von Si und β-SiC bzw. die (0001), (10$\bar{1}$0) und (12$\bar{1}$0) im Falle von α-Si$_3$N$_4$ Einkristallen durchgeführt. Auf diese Weise konnten sowohl die Härten und die Elastizitätsmoduli von verschiedenen Materialien ermittelt werden, als auch die Ergebnisse der MD-Simulation in eine Beziehung mit den experimentellen Untersuchungen der mechanischen Eigenschafften von Si, SiC und Si$_3$N$_4$ gebracht werden.

In **Kapitel 11** wird die Studie der Nanoindentation an SiC- und Si$_3$N$_4$-Schichten auf Siliziumsubstraten vorgestellt. Der Beschichtungsprozess an sich wurde bereits in Kapitel 9 beschrieben, während die Härtemessungen an Einkristallen in **Kapitel 10** stattfanden. Für die Untersuchung der mechanischen Eigenschaften von Schichtsystemen wurden stöchiometrische SiC- und Si$_3$N$_4$-Schichten verwendet, während die Nanoindentation von Siliziumschichten und gradierten SiC- und SiN-Schichten als eine mögliche Fortsetzung der Untersuchungen anzusehen ist. Molekulardynamiksimulationen der Nanoindentation wurden mit experimentellen Messungen unter Verwendung eines Nanoindenters UMIS 2000 verglichen und validiert. Die experimentellen Untersuchungen fanden am IAM-AWP des Karlsruhers Instituts für Technologie statt. Die Ergebnisse dieser Studie, die gemessenen Härten sowie der Substrateinfluss auf die ermittelte Schichthärte, sind auch in [18, 71] zu finden. Eine Zusammenfassung der wichtigsten Ergebnisse dieser Arbeit kann in **Kapitel 12** gefunden werden.

Die Ergebnisse der vorliegenden Arbeit wurden mehrfach auf internationalen Konferenzen vorgestellt und resultierten in insgesamt drei Veröffentlichungen. Eine Liste von Vorträgen und Veröffentlichungen kann in **Kapitel 13** eingesehen werden. Im Laufe der Arbeiten am IMWF konnte zudem ein Beitrag für die Lehre geleistet werden: So konnte z. B. die Anfertigung einer Studienarbeit erfolgreich betreut werden, während sich eine Masterarbeit noch in Bearbeitung befindet. Nicht zuletzt sind Ergebnisse dieser Arbeit in Vorlesungen an den Universitäten Stuttgart (Prof. Dr. Siegfried Schmauder) und Karlsruhe (PD Dr. Sven Ulrich) eingeflossen.

3 Introduction

3.1 Computational Materials Science

During the last decades, aside to classical methods of theory and experiment, computer simulations have emerged as a very powerful method for the description of physical reality. On the one hand they are based on theoretical models which are solved numerically, on the other hand their ability to reproduce an experimental situation has also made them very popular to experimental scientists. It is the deterministic nature of a computer simulation which makes it possible to reproduce a result, assumed the parameters of the computer simulation remain unchanged, which is a cornerstone of the experiment as a scientific method, making the computer simulation in a way to a *super microscope* [88]. A scientist can change parameters of the simulations independently and observe their effect onto simulation results, something which is not always possible in a real experiment, due to the boundary conditions set by the used apparatus.

Depending on the typical length scale of the process, different computer models can be used, describing the physical situation by appropriate variables. Starting from the lowest length scale, density functional theory (DFT), Quantum Monte Carlo (QMC) [129] and Hartree Fock Finite Element Method (HFFEM) [17, 35, 98, 115] are used to describe the behavior of electrons and of the atomic nuclei. With increasing length scale, the exact knowledge of the electronic state vector at every time point becomes negligible and time- and space averages of individual electronic states can be used instead.

The time development of a system of atoms is governed by present

Chapter 3. Introduction

interatomic forces, individual electrons are not represented directly but rather in the form of effective interatomic potentials. On this scale, ranging from one Ångstroem up to several micro-meters in length and from one femtosecond to several nanoseconds in time, molecular dynamics (MD) is the most commonly used method. In the framework of MD, atoms are represented as mass points, their movement can solely be described by Newtonian mechanics. The remaining variable required for the description of an N-particle system governed by classical mechanics is the force acting on individual particles, it is represented in the form of parameterized potential functions which depend on the distance, angle and/or local environment.

One of the most simple potentials is the Lennard-Jones (LJ) pair potential [82, 83] which depends only on the distance between atoms and aims for representing the weak van der Waals forces acting between neutral atoms such as noble gases. In case of metals, embedded atom method (EAM) [106, 138] is used for the description of the interaction between individual positively charged metal ions as well as of the interaction between the ions and delocalized electrons surrounding them. Covalent materials such as diamond and silicon are represented by so called bond-order-potential (BOP) functions [45, 74, 75, 76, 77, 78, 89, 96]. The strong angular dependency is one of the main characteristics of these potentials, required for the representation of the directional nature of covalent bonds. In addition, the attractive term of the potential function is also dependent on the local enviroment: the more atoms within the vicinity of a specific atom, the weaker the bond to the individual atom. This feature is required for the description of bond hybridization such as sp^2 and sp^3 which occurs when two hybridization states are separated by a low energy barrier. Carbon, in its diamond and graphite structure is one of the most prominent examples of the hybridization and was the element which initiated the development of bond-order potentials [75]. A more complex material, which will be described in this work is silicon carbide, crystallizing in both hexagonal α-SiC and cubic β-SiC phase, see **Figures 3.4(a)** and **3.4(b)**.

Section 3.1. Computational Materials Science

Molecular dynamics is capable of describing a large number of atoms e.g. 10^9 for a period of time up to several nanoseconds. However, a quick calculation reveals the limitations of this method: macroscopic problems deal with numbers of atoms of around 10^{23} for time periods ranging to several years. Therefore, further simulation methods are required for a suitable description of these processes: dislocation dynamics (DD) [12, 145] and phase field method (PFM) [111] are interesting at longer time and longer length scales. Processes on these scales are governed by thermodynamic forces, trajectories of individual atoms become obsolete, moreover an average over a large number of particles, called ensemble, is made. Dislocation dynamics and phase field methods are often referred to as intermediate scale or mesoscale methods, having the previously mentioned methods as a lower scale and the finite element method (FEM) as the upper scale.

The finite element method (FEM) originated from the need for solving complex elasticity and structural analysis problems in civil and aeronautical engineering [2]. Within the FEM frame, a macroscopic sample is divided or meshed into a large number of well defined volumes or finite elements. For each element, differential equations are solved independently, taking care of the function steadiness and differentiability at element interfaces.

All of the methods mentioned previously are several decades old, a cross-linked usage of these methods unified by the idea of modeling the physical reality by the means of numerical simulation has emerged to a unified concept of Multiscale Materials Modeling (MMM) [21, 139], see **Figures 3.1** and **3.2**. Within the concept of MMM, every problem is addressed at the appropriate length scale, the link between individual scales can be a direct, horizontal approach, or an indirect, vertical approach. In the horizontal approach, several scales are used in parallel, a system of interest is decomposed into regions, within each region a well defined scale is used. The simulation evolves by solving differential equations within each scale as well as by the movement of inter-scale boundaries. The probably best example for this approach is the model-

Chapter 3. Introduction

ing of crack propagation, presented in **Figure 3.1**, source [146]. Regions at larger distance from the crack tip are modeled by the finite element method (FEM) while molecular dynamics is used to model regions closer to the crack tip. In the vicinity of the crack tip, a quantum mechanical approach is adopted, representing the crack propagation as a continuous breaking of atomic bonds. One big advantage of this approach is surely the high accuracy that can be reached by relatively small computation times, since only a smaller part of the system has to be modeled by time consuming methods such as MD and DFT.

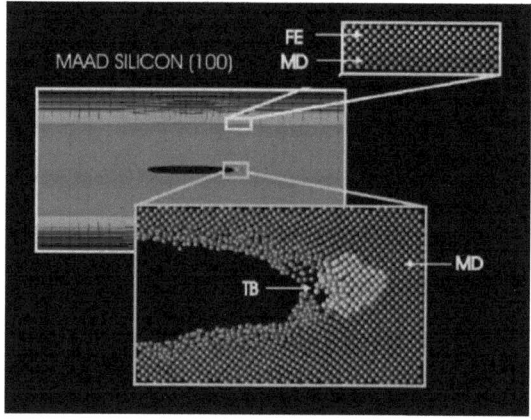

Figure 3.1: Multiscale simulation of a crack propagation in silicon, source [146]. Crack propagation is a standard example for the horizontal approach in Multiscale Materials modeling (MMM). While the outer regions of the sample are modeled by Finite Element Method (FEM), the regions closer to the crack tip (here in blue color) are modeled by Molecular Dynamics (MD). In the vicinity of the crack tip itself, further simulation methods, such as Density Functional Theory (DFT) can be applied or, like in this example, chemical bonds in silicon are presented as Tightly Bound states (TB).

Section 3.1. Computational Materials Science

The biggest disadvantage of the horizontal approach is the correct mathematical description of the interface between two neighbouring methods. It is mostly because of this issue why the horizontal approach in Multiscale Materials Modeling is not so wide-spread as one would expect, considering that all individual methods are well established. The vertical approach in MMM circumvents this problem by treating each scale in the appropriate manner and bridging them by a stepwise parameter transfer from the lower scale to the upper one. In this way the problem is treated in a serial, rather then parallel manner. **Figure 3.2** presents methods used in materials modeling together with their individual length scales and problems which are adressed by individual methods. Some other interesting contributions to Multiscale Materials Modeling can be found in [21, 139].

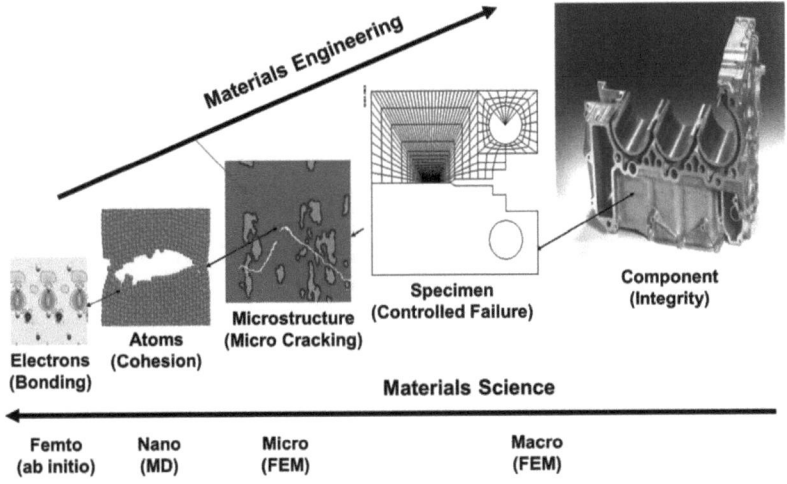

Figure 3.2: Length scales in Multiscale Materials Modeling (MMM). By a stepwise parameter transfer from a lower scale to the upper one, it is possible to describe the whole range of material properties.

Chapter 3. Introduction

An important example of the vertical approach is the modeling of the precipitate-dislocation interaction. Kinetic Monte-Carlo is a simulation method which aims to describe the ageing process of steels using a fixed crystal lattice and modeling the diffusion process of individual impurity atoms as a constant movement of a single vacancy site in the crystal. This makes it possible to describe the growth process of Cu/Ni/Mn precipitates in an alloyed iron.

By molecular dynamics, the critical resolved shear stress in dependence of the precipitate size and composition can be calculated, hereby quantifying the strengthening mechanism. Precipitate size distribution in a large system can be calculated by the phase field method (PFM).

Detailed description of the exact procedure can be found in a publication within the work group at IMWF made by Molnar et al. [39]. All variables known are calculated by simulation methods at different time and length scales, allowing one to make statement about macroscopic observables such as strength, hardness and material stiffness.

3.2 Description of the material system

In **Figure 3.3**, a ternary Si-C-N material system is presented. Within it, two thermodynamic stable binary alloys can be found, namely SiC and Si_3N_4. In addition, the position of the theoretically predicted metastable C_3N_4 phase is marked [22]. Together with BN and B_4C, these are the most important non-oxide ceramics currently used. Amorphous ternary alloys, such as a-Si-C-N [108, 109] and a-Si-B-N [32, 97] are also used and/or are a subject of intensive research. **Table 3.1** presents an overview of basic material properties of silicon and two different polytypes of SiC and Si_3N_4.

Section 3.2. Description of the material system

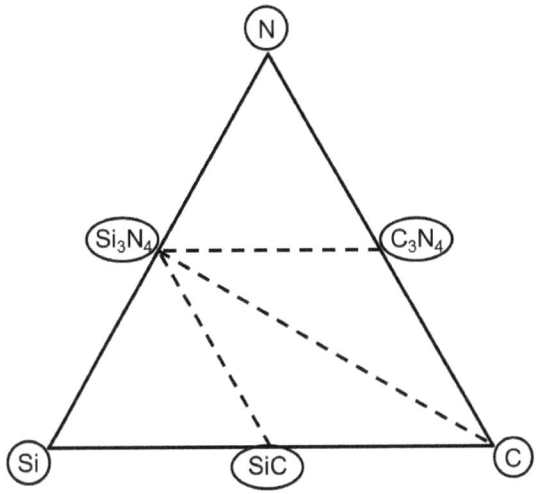

Figure 3.3: Schematic representation of the Si-C-N material system.

Material	Silicon	α-SiC	β-SiC	α-Si$_3$N$_4$	β-Si$_3$N$_4$
Structure	diamond	hexag.	zincblende	trigonal	hexag.
Latt. const. a	5.431	3.073	4.32	7.66	7.586
Latt. const. c		10.053		5.615	2.902
Density	2.329	3.21	3.166	3.18	3.44
Bind. energy	4.61	9.52	6.17	11.28	
Covalency	1.0	0.998	0.998	0.739	0.739
Mechanical properties					
Vickers hardness	1000	3500	2400-3300	1400-3200	2500
Young's mod.	130-180	440	480	320	290

Table 3.1: Material properties of silicon compared with SiC and Si$_3$N$_4$, source [43, 99]. Lattice constants a given in [Å], density in [g/cm^3], binding energy in [eV] and Young's modulis in [GPa].

Chapter 3. Introduction

3.2.1 Silicon carbide - SiC

Silicon carbide (SiC) is a semiconductor material with desirable properties for many applications, primarily due to its wide energy band gap (3.0-3.2 eV) and secondly due to its high thermal and mechanical stability. Si, GaP and GaAs are commonly used wide band gap semiconductor materials for electronic devices. All three materials, however, have much lower melting temperatures than SiC, making SiC the perfect candidate for high-temperature electronic components. This, together with other important properties of SiC as a semiconductor in comparison with Si, GaP and GaAs are given in **Table 3.2**.

Property	Si	GaAs	GaP	β-SiC	Diamond
Band gap [eV]	1.10	1.40	2.30	2.20	5.50
Max. oper. temp. [K]	600	760	1250	1200	1400
Melting point [K]	1690	1510	1740	2100	phase change
Physical stability	good	fair	fair	excellent	very good
Electron mobility	1400	8500	350	1000	2200
Vacancy mobility	600	400	100	40	1600
Thermal conductivity	1.5	0.5	0.8	5	20
Dielectric constant	11.8	12.8	11.1	9.7	5.5

Table 3.2: Comparison of some relevant properties between SiC and other important wide-gap semiconductor materials, source [126]. Electron and vacancy mobility are given in [cm^2/Vs] while thermal conductivity is in [WK/cm].

Silicon carbide appears in two crystal modifications, in a cubic or β-SiC phase, stable at low temperature and low pressure and different hexagonal and rhombhoedral crystal structures or α-SiC, which form from β-SiC above 1000 °C [59]. In dependence on the orientation and the stacking sequence of individual tetrahedra layers, different polytypes of α-SiC can be distinguished, such as 4-H-, 6-H- and 15-R-α-SiC [99]. The crystal structure of the α- and β-phase of SiC is presented in **Figure 3.4**. Special emphasis in this work will be given to diamond-like β-SiC.

Section 3.2. Description of the material system

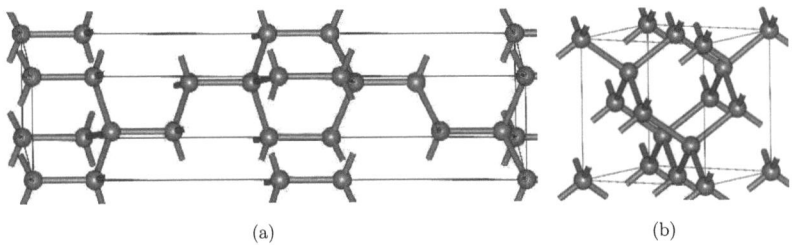

(a) (b)

Figure 3.4: Crystal structure of the α- (a) and β-SiC phase (b), source [7]. Both silicon and carbon atoms are shown in grey color, as only Si-C bonds are present and the stoichiometry is Si_1C_1, the crystal lattice is symmetric towards the exchange between silicon and carbon atoms.

Next to its desirable properties as a semiconductor, high mechanical stability, good corrosion resistance as well as high thermal stability are also properties associated with silicon carbide. SiC can absorb and emit light in the whole visible frequency range, making it ideal for the application in color displays. It is also possible to use SiC for the protection of nuclear plant walls for shielding purposes against tritium.

3.2.2 Silicon nitride - Si_3N_4

Silicon nitride, Si_3N_4, is normally not regarded as a semiconductor, however its large band gap of 5 eV makes it a valuable candidate for isolation and passivation coatings in the semiconductor industry [99, 127], antireflexion coatings for solar cells [24], as well as oxidations masks and diffusions barriers [30, 42, 60, 64, 85, 143]. In addition, its high thermal, mechanical and corrosion stability makes it an interesting material for steel protective coatings, X-ray masks and protection of thermonuclear reactor walls.

Si_3N_4 and SiC show a large amount of similarity when it comes to possible application in engineering, e.g. protection of structural parts in gas turbines, catalytic heat exchangers and combustion systems. **Table 3.1** summarizes the most important material properties of Si_3N_4, together with SiC and Si.

Chapter 3. Introduction

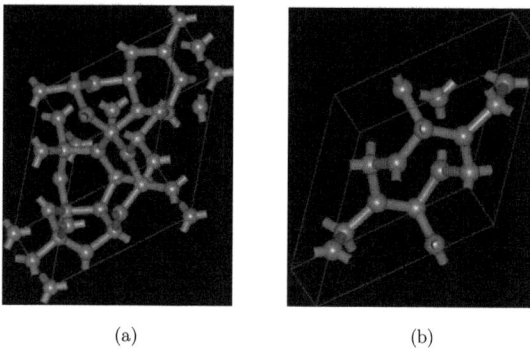

(a)   (b)

Figure 3.5: Crystal structure of the α- (a) and β-Si$_3$N$_4$ phase (b), silicon atoms are presented in gray, while blue atoms represent nitrogen, source [8].

Si$_3$N$_4$ is present in two modifications, having the same stoichiometry, but different crystal structures. At low temperature and low pressure, it appears either in the metastable, trigonal α-Si$_3$N$_4$ or the stable hexagonal β-Si$_3$N$_4$ modification. Experimentally, Si$_3$N$_4$ can be deposited by chemical (CVD) or physical (PDV) vapor deposition method at high substrate temperatures (above 700 °C). The structure and the stoichiometry of these coatings depends highly on process parameters and is predominantly amorphous, with hardnesess between 35 GPa and 56 GPa [149].

4 Molecular Dynamics (MD) method

While a short overview onto a variety of computer simulation methods was given in the previous chapter, a more detailed insight into the method of molecular dynamics will be provided here.

4.1 Quantum-mechanical approach

In general, materials can be considered as a large number of interacting atoms, their physical and chemical behavior is solely determined by the electro-magnetic force acting between atoms. An individual atom can be further subdivided into the positively charged nucleus and negatively charged electrons, the description of the system at this level can be done by solving the many-body Schrödinger equation 4.1. The evolution of an N-particle system is described by the action of the Hamilton operator $\hat{H}$ onto the wave function Ψ. Within this frame, the state of the system containing N interacting particles is represented in the form of a wave function Ψ depending on $3N$ space coordinates, its modulus $|\Psi|$ being the probability density. The Hamilton operator $\hat{H}$ consists of the kinetic energy of individual nuclei T_N, the kinetic energy of the electrons T_e as well as the nuclei-nuclei V_{NN}, electron-nuclei V_{eN} and the electron-electron V_{ee} interaction, as given through the equation:

$$\hat{H}\Psi = E\Psi \quad (4.1)$$

Chapter 4. Molecular Dynamics (MD) method

$$[T_N + T_e + V_{ee}(r) + V_{NN}(R) + V_{eN}(r,R)]\Psi = E\Psi \qquad (4.2)$$

Since all of the acting particles are fermions, the exclusion principle leads to the analytical representation of the N-particle wave function Ψ in the form of a linear combination of individual wave functions $\psi_i(\vec{r}_i)$, one of them being the Slater determinant:

$$\Psi = \frac{1}{\sqrt{N!}} \begin{vmatrix} \psi_1(\vec{r}_1) & \cdots & \psi_1(\vec{r}_N) \\ \vdots & & \vdots \\ \psi_N(\vec{r}_1) & \cdots & \psi_N(\vec{r}_N) \end{vmatrix} \qquad (4.3)$$

Due to their low mass, it can be assumed that electrons instantly follow every movement of the nuclei, simplifying the state representation by separating the dynamics of the nuclei, presented by the wave function $\Psi_N(R)$ and of electrons, presented by the wave function $\Psi_e(r,R)$:

$$\Psi_{atom}(r,R) = \Psi_e(r,R) \cdot \Psi_N(R) \qquad (4.4)$$

$$[T_e + V_{ee}(r) + V_{eN}(r,R)]\Psi_e = \varepsilon(R)\Psi_e \qquad (4.5)$$

$$[T_N + V_{NN}(R) + \epsilon(R)]\Psi_N(R) = E_N\Psi_N \qquad (4.6)$$

The electronic eigenvalues are therefore only parametrically dependent on the nuclei positions. The Schrödinger equation 4.6 of the nucleus describes its behavior in a potential of surrounding nuclei $V_{NN}(R)$ and the electronic embedding potential $\epsilon(R)$. This approximation is formally known as the Born-Oppenheimer approximation and with few exceptions of very light nuclei such as H$^+$ and He^{2+} it can be taken as justified for all further considerations. While the electron dynamics remains in the domain of the quantum mechanics, the classical approximation can be used for the treatment of the nuclei, allowing one to use Newtonian mechanics for system description. It is exactly here where the method of molecular dynamics originates from. The dynamics of the electronic system is used indirectly, in the form of interatomic potentials, which can have different forms, depending on the material system under investigation. While interatomic forces are of quantum mechanical origin, the remaining part of the equation 4.1 is a set of $3N$ coupled differential equations requiring only newtonian treatement.

4.2 Differential equations

For a system of N interacting particles, the time development of the system can be described by $3N$ coupled differential equations of the form:

$$-\vec{\nabla} V(r_i) = m \frac{\partial^2 r_i}{\partial t^2} \qquad (4.7)$$

$$r_i = (x_i, y_i, z_i) \qquad (4.8)$$

where r_i is a $3N$-dimensional cartesian vector representing the coordinates of individual particles and $V(r)$ is the interatomic potential. The gradient of the interatomic potential $V(r)$ represents the force acting on individual particles and resulting into the acceleration given in the form of the second time derivative of the position vector r_i.

This set of coupled partial differential equations can not be solved in an analytical manner, however numerical integration can be performed, using Δt as an integration step. The convergence of the result delivered by the numerical integration depends highly on the appropriate choice of the integration step Δt, the smaller the value of Δt, the closer is the result to the analytical solution. At the same time, smaller time steps increase the computation time. The proper choice of the integration time is, therefore, crucial for the convergence of the result and has to be adapted individually to the problem under investigation. Iterating over time steps, initial conditions such as atomic coordinates, velocities and potential energies are transferred from one time step to the next one and the coupled differential equations of nuclei motion are recalculated.

Chapter 4. Molecular Dynamics (MD) method

This approach makes it necessary to have fast algorithms capable of solving differential equations.

4.3 Algorithms for numerical integration

In following sections, a short overview across algorithms used for numerical integration will be given. These algorithms are standard tools for numerical integration of the equations of motion, presented in the previous section, and are implemented in many other molecular dynamics codes such as **ESPResSO** [9] or **LAMMPS** [10]. The knowledge of basic features of numerical integrators is essential for a better unterstanding of the molecular dynamics method itsself. Another examples of a summary of algorithms used in MD can be found in [46, 130].

4.3.1 Verlet algorithm

The Verlet algorithm uses the sum of two symmetrical Taylor expansions of $r(t + \Delta t)$ and $r(t - \Delta t)$:

$$r_i(t + \Delta t) = r_i(t) + \Delta t \dot{r}_i(t) + \frac{1}{2}(\Delta t)^2 \ddot{r}_i(t) + ... \qquad (4.9)$$

$$r_i(t - \Delta t) = r_i(t) - \Delta t \dot{r}_i(t) + \frac{1}{2}(\Delta t)^2 \ddot{r}_i(t) - ... \qquad (4.10)$$

Due to this sum, the odd powers of Δt are substracted and velocities do not appear explicitly:

$$r_i(t + \Delta t) = 2r_i(t) - r_i(t - \Delta t) + (\Delta t)^2 \ddot{r}_i(t) + ... \qquad (4.11)$$

Force F is represented as the gradient of the interatomic potential V acting on an atom with a mass m and resulting in an acceleration $\ddot{r}$:

$$\ddot{r} = \frac{F}{m} = -\frac{1}{m}\frac{\partial V}{\partial r} \qquad (4.12)$$

Moreover, the velocity can be calculated from the subtraction:

Section 4.3. Algorithms for numerical integration

$$\dot{r}_i(t) = \frac{r_i(t+\Delta t) - r_i(t-\Delta t)}{2\Delta t} \tag{4.13}$$

Due to the symmetry of the Verlet algorithm, it is time-reversible, in sum it is correct up to the order Δt^4. Using the Verlet algorithm for numerical integration guarantees energy conservation for small time steps Δt such as they are used in molecular dynamics.

4.3.2 Leapfrog algorithm

An improvement of the Verlet algorithm is presented with the Leapfrog algorithm, which uses velocities as an additional degree of freedom, rather than calculating them from positions, as in equation 4.13. In this way, terms of order $O^2(\Delta t)$ of the time step are avoided, allowing higher numerical precision. The methodology of the Leapfrog algorithm becomes evident from equations 4.14 and 4.15: rather then having one differential equation of the second order, two differential equations of the first order are calculated. Calculation of velocities $\dot{r}_i(t+\frac{\Delta t}{2})$ is shifted by the size of $\frac{\Delta t}{2}$ compared to the calculation of positions r_i:

$$r_i(t+\Delta t) = r_i(t) + \Delta t \dot{r}_i\left(t+\frac{\Delta t}{2}\right) \tag{4.14}$$

$$\dot{r}_i\left(t+\frac{\Delta t}{2}\right) = \dot{r}_i\left(t-\frac{\Delta t}{2}\right) + \Delta t \ddot{r}_i(t) \tag{4.15}$$

In analogy to equation 4.13, the velocity $\dot{r}_i(t)$ at the time step t can be calculated, if necessary, by averaging over velocities at two subsequent time steps, according to equation:

$$\dot{r}_i(t) = \frac{\dot{r}_i(t+\frac{\Delta t}{2}) + \dot{r}_i(t-\frac{\Delta t}{2})}{2} \tag{4.16}$$

4.3.3 Velocity-Verlet algorithm

The Leapfrog algorithm shows significant benefits regarding precision, compared with the Verlet algorithm. It is in many cases necessary in MD to know velocities and positions for the same integration step, e.g. for the calculation of stresses. In order to enable this, Velocity-Verlet algorithm is used, which allows the calculation of the velocities directly out of forces:

$$r_i(t + \Delta t) = r_i(t) + \Delta t \dot{r}_i(t) + \frac{1}{2}(\Delta t)^2 \ddot{r}_i(t) \qquad (4.17)$$

$$\dot{r}_i(t + \Delta t) = \dot{r}_i(t) + \frac{\Delta t}{2}[\ddot{r}_i(t) + \ddot{r}_i(t + \Delta t)] \qquad (4.18)$$

4.4 Thermodynamic ensembles

Using the Verlet algorithm (equations 4.9 and 4.10), Leapfrog algorithm (equations 4.14 and 4.15) or Velocity-Verlet algorithm (equations 4.17 and 4.18) allows the description of the time development of a closed system with a constant number of particles, volume and energy. In order to introduce temperature and/or pressure control, required for the simulation of more realistic systems, equations 4.9-4.18 have to be suitably modified.

4.4.1 Temperature control

An additional friction term has to introduced in the equations of motion 4.9-4.18 in order to rescale the velocities and enable heating or cooling of the system. The Nose-Hoover thermostat [133] correctly reproduces the isothermal-isobaric partition function and is, therefore, used here. With $\dot{r}_i$, p_i and m_i being particle velocity, particle impulse and particle mass, modified equations of motion for an N-particle system in D dimensions with N_D degrees of freedom contained in a volume V are then:

Section 4.4. Thermodynamic ensembles

$$\dot{r}_i = \frac{p_i}{m_i} \tag{4.19}$$

$$\dot{p}_i = F_i - \eta p_i \tag{4.20}$$

Particle impulse p_i is, therefore, not only influenced by the force F_i, equation 4.12, acting on the particle but is furthermore multiplied by the friction factor η, defined in the equation 4.20. The friction factor η itself depends on the ratio of the current system temperature T_{current} and the temperature of the heat bath T_{desired}, k is the Boltzmann constant and τ_T is the characteristic thermostat relaxation time. During the simulation, η changes according to equation 4.21, where the system temperature T_{curent} is calculated from the average particle kinetic energy, according to equation 4.22:

$$\dot{\eta} = \frac{1}{\tau_T^2}\left(\frac{T_{\text{current}}}{T_{\text{desired}}} - 1\right) \tag{4.21}$$

$$T_{\text{current}} = \frac{1}{3Nk}\sum_i \frac{|p_i|^2}{m_i} \tag{4.22}$$

4.4.2 Pressure control

Next to temperature control, pressure control has to be established in order to accurately reproduce the gross-canonical ensemble (NPT). In analogy to the temperature control, the pressure control depends on the current system pressure p_{current} and the desired pressure p_{desired}, the system volume V is rescaled at every time step in order to accomodate the pressure. Theoretical background of the implementation of the grand canonical ensemble (NPT) in molecular dynamics can be found in [131]. The pressure p_{current} of an N-particle system in a volume V is calculated from particle impulses p_i and forces acting between particles F_{ij} with the distance vector $r_i - r_j$:

Chapter 4. Molecular Dynamics (MD) method

$$p_{\text{current}} = \frac{1}{3V}\left(\sum_i \frac{|p_i^2|}{m_i} + \frac{1}{2}\sum_{i,j} r_{ij} \cdot F_{ij}\right) \quad (4.23)$$

An additional parameter ξ is introduced acting both on particle velocities $\dot{r}_i$ and impulses p_i as well as on the system dimensions h as presented in equations 4.24-4.26. Depending on system pressure, system volume is rescaled in order to increase or decrease the internal pressure. The volume rescaling is either isotropic, rescaling all system dimensions equivalently, or axial, rescaling each axis of the system independently:

$$\dot{r}_i = \frac{p_i}{m_i} + \xi r_i \quad (4.24)$$

$$\dot{p}_i = F_i - (\xi + \eta)p_i \quad (4.25)$$

$$\dot{h} = \xi h \quad (4.26)$$

Parameter ξ is varied during the simulation in dependence on the temperature T_{current}, the current system pressure p_{current} and desired pressure p_{desired}:

$$\dot{\xi} = \frac{1}{NkT_{\text{current}}\tau_p^2}V(p_{\text{current}} - p_{\text{desired}}) \quad (4.27)$$

Parameter τ_p (similar to τ_T in equation 4.21) is the characteristic pressure relaxation time and has to be chosen appropriately for the system under investigation.

As stated in the beginning of this chapter, molecular dynamics is also suited for the investigations of systems close to the thermodynamic ground state. An example of possible usage of this aspect is the rapid quenching of a molten phase in order to produce a homogeneous, amorphous phase of a system under consideration. In general, two relaxators are implemented in most MD codes, following the same idea, but performing slightly different.

4.4.3 Micro convergence integrator

A microconvergence integrator (MIK) is a numerical integrator which resets the impulse of a particle to zero if the scalar product of the particle impulse and the force acting on the particle is negative, or simplifying the statement, the particle goes "uphill" in the potential landscape, [73].

4.4.4 Global convergence integrator

A globalised version of the microconvergence integrator is GLOK. Per definition, GLOK resets the impulse of all particles to zero if the scalar product of the global impulse (sum of individual particle impulses p_i) and the global force (sum of individual forces F_{ij} acting between the particles) acting on the atoms is negative, [73].

All numerical integrators described at this stage are standard tools in molecular dynamics and are used in various simulation codes. Within this work, the micro canonical ensemble (NVE) was used for the modelling of the sputtering process. Upon an impact of an Ar^+ ion on a target surface, all processes within the collision cascade are fast (few hunderd fs) and therefore occur far from thermodynamics equilibrium in a so-called *thermal spike* regime, see **Chapters 5** and **7**. A good example for the usage of the temperature and pressure control is the modelling of the deposition process, where the substrate is in direct contact with its surroundings. In experiment often an additional heating of the substrate is necessary in order to deposit nanocrystalline coatings. Relaxation integrators are the method of choice for the simulation of the nanoindentation, since the dislocation formation, their propagation and the exact measurement of the critical resolved shear stress required to initiate dislocation propagation can be easily measured at 0 K. At higher temperatures, these effects are superimposed by thermal noise, which may lead to their underestimation.

Chapter 4. Molecular Dynamics (MD) method

4.5 Periodic boundary conditions

The concept of periodic boundary conditions (PBC-s) is often related to molecular dynamics. PBC-s are used to represent an infinite volume and number of particles by calculating only a finite number of particles placed in a certain *unit volume*. This unit volume is then periodically repeated to the positive and negative side of all three axes of the cartesian coordinate system. Peridodic boundary conditions were introduced for the first time by Max Born and Theodore von Karman in 1912 [95] and are described in more detail in [105]. The usage of the periodic boundary conditions is required in order to reduce or completely remove surface effects which would normally lead to their overestimation in an MD simulation, due to the fact that the volume/surface ratio in the simulated system is by several orders of magnitude smaller than in reality.

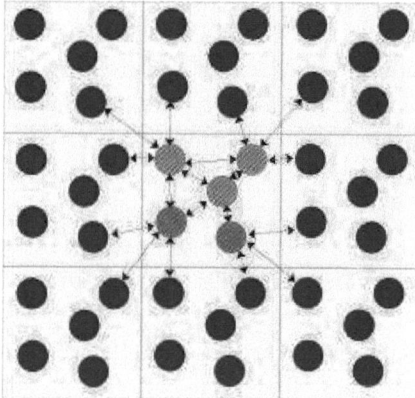

Figure 4.1: Schematic representation of a two-dimensional periodic system. Particles can move within the system, if a particle leaves the system across one of the four edges, it will enter back into the system trough the opposite edge. In three dimensions the four edges are replaced by six cube faces [95, 105].

4.6 Theory of interatomic potentials

Classical interatomic potentials are designed for the modeling of interactions between individual mass points. As described in previous sections, solving the electronic part of the Schrödinger equation is not the assignment of the molecular dynamics method, however, the exact knowledge of the solution is the crucial ingredient of a realistic molecular dynamics simulation. Because of this, the solution of the Schrödinger equation is presented in the form of a parameterized function or a potential. The functional form of the potential as well as the parameters within an individual functional form are material dependent, their purpose is to represent as many material properties as possible. In general, having a specific functional form of a potential, it is possible to adjust the parameters and fit the potential to macroscopic observables such as the lattice constant, elastic constants or melting temperature. Potentials developed in this way are often referred to as empirical potentials. Another, more exact, but also more demanding possibility of potential development is based on density functional theory (DFT) or ab initio method. Both methods have minor differences, but their essential feature is that the potential parameters are based completely on a quantum mechanical calculation, without any empirical values.

It must be stated at this point that it is almost impossible to represent every aspect of a specific material by one single potential, moreover, many potentials can coexist for one single material. The probably best example for the coexistence of several interatomic potential describing the same material system is silicon, the fact that it is a semiconductor makes the necessity of different potential forms obvious, some of them, such as Stillinger-Weber [48] and Tersoff potential [74, 76, 77] describe its covalent nature much better, while others, such as Embedded Atom potential EAM [102] are better in representing its metallic properties. At last there exists the tendency to have one potential for the description of all silicon properties, both its covalent nature with angular dependent chemical bonds as well as its metallic nature, e.g. Modified Embedded

Chapter 4. Molecular Dynamics (MD) method

Atom Method MEAM [103, 104] or the potential of Albe [112]. For other material systems also several potentials can coexist, each of them describing a certain number of material properties differently well.

Separating the nuclei part of the Schrödinger equation 4.1 and applying Born-Oppenheimer approximation, equation 4.4, the potential energy of N particles having r_n ($n = 1...N$) for coordinates can be described as a series of single-, two-, three- and many-body terms:

$$V = V_1 + V_2 + V_3 + ... \qquad (4.28)$$

The first term in the series is only position dependent and equals zero in absence of external forces simplifying the expression:

$$V(r_1, ..., r_N) = \sum_{i<j}^{N} V_2(r_i, r_j) + \sum_{i<j<k}^{N} V_3(r_i, r_j, r_k) + ... \qquad (4.29)$$

The convergence of the equation 4.29 depends on the material system under consideration, e.g., for the representation of noble gases, the summation up to the two body part is sufficient. Metals and covalent elements form complex crystal structures with angular dependent bonds of large range and, therefore, require taking into account the three body part or even higher terms.

4.6.1 Pair potentials

The most simple form of potentials represent the pair potentials. They depend only on particle distance, their range as well as the order of the distance dependency reflect the physical nature of the chemical bond. In the following, three examples for pair potentials will be given.

Ionic bond

Sodium chloride or common salt is probably the best example for an ionic solid. Due to a large difference in electronegativity, a complete charge transfer from Na towards Cl atom occurs. Positively charged

Section 4.6. Theory of interatomic potentials

Na$^+$ and negatively charged Cl$^-$ ions of the rock-salt structure attract each other while repulsing ions of the same electronic charge, interacting via Coulomb pair potential.

Van der Waals bond

Simple pair potentials are also used for the numerical representation of interatomic forces in case of noble gases. The nature of the interaction is, however, more complicated and is based on dipole-dipole interaction of neutrally charged atoms, it is only the inhomogeneity of the electronic charge distribution for a short period of time which results in the attraction of two atoms. Therefore, the potential describing van der Waals forces is an attractive central potential depending only on the distance between individual atoms, the attraction is of the order $O(r^{-6})$. In addition to the attractive part of the van der Waals bond, the interaction has to be stabilized by a repulsive part of higher order, $O(r^{-12})$, in order to avoid zero distances and potential singularity. Physically, the introduced repulsive potential part represents the repulsion between two neutral atoms caused by overlapping of closed electronic shells of individual atoms. Therefore, the pair potential representing van der Waals forces can be written as:

$$V_{ij} = -4\epsilon \left[\left(\frac{r_0}{r_{ij}}\right)^6 - \left(\frac{r_0}{r_{ij}}\right)^{12} \right] \quad (4.30)$$

This potential is formally known as the Lennard-Jones LJ potential, the parameters ϵ and r_0 represent the strength of the bond and the radius of the repulsive core, their values are material dependent.

Ziegler-Biersack-Littmark potential

Another type of pair potentials are pure repulsive potentials. One of the most prominent potentials of this class is surely the potential developed by Ziegler, Biersack and Littmark, better known in its abbreviation form as the ZBL-potential [155] .

Chapter 4. Molecular Dynamics (MD) method

The ZBL-potential was developed to model repulsive interactions such as they are present in ion radiation damage in nuclear power plants or in radiation experiments. Within such processes kinetic energies of incident ions are often in the range up to several 10^8 eV, in dependence of the ion origin process. Due to ion radiation, collision cascades are formed as the ion penetrates the material, where electronic shells of the ion and material atoms overlap in high range, in sum, also the repulsive force in between the two nuclei is included in the ZBL-potential. At these small distances it is required to have a repulsive potential, which decreases fast enough as the distance of the particles increases. The ZBL-potential is a screening electrostatic potential for nucleus-nucleus interaction and possesses the following form:

$$V(r) = \frac{Z_1 Z_2 e^2}{r} \phi(r) \qquad (4.31)$$

$$\phi(r) = \sum_{1,4} A_i \exp\left(-b_i \frac{r}{a_u}\right) \qquad (4.32)$$

$$a_u = \frac{0.8854}{Z_1^{0.23} + Z_2^{0.23}} a_{\text{Bohr}} \qquad (4.33)$$

Z_1 and Z_2 represent atomic numbers of the interacting particles at a distance r, e is the electron unit charge and $\phi(r)$ is the screening function defined in equation 4.32. The screening length a_u is defined in the Eq. 4.33 where a_{Bohr}=0.529 Å is the Bohr radius.

Indenter-substrate potential

Nanoindentation is a method commonly used for the investigation of mechanical properties. In general, a hard indenter tip is pushed in a well controlled manner in the material under investigation while measuring external load and penetration depth. Experimentally, different hard materials, such as diamond or silicon nitride are used for indenter tips. Numerically, the indenter is represented as a rigid body, therefore, the description of interactions within the indenter is redundant.

Section 4.6. Theory of interatomic potentials

For the simulation of the interaction between the rigid indenter and the material under investigation commonly different potentials, such as ZBL-potential [84] or Morse potential [122] are used.

Within the scope of this work, an own potential was developed and applied. The potential form is given in equation 4.34 while a graphical representation is shown in **Figure 4.2**.

$$V(r) = 0.1 \cdot \frac{1}{(r-r_0)^{16}} \quad (4.34)$$

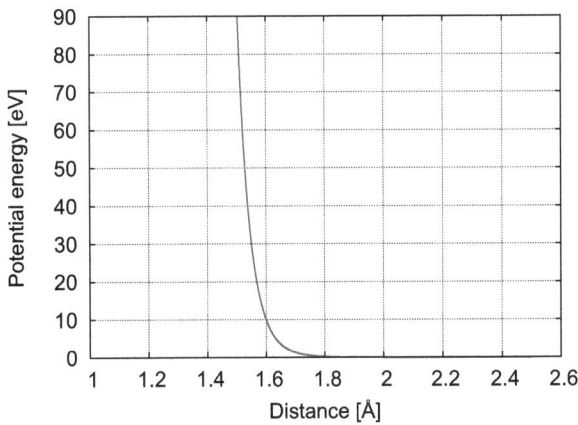

Figure 4.2: Steric pair potential used for the indenter-substrate material interaction, described in equation 4.34.

The potential developed here is a pure repulsive potential with a high order distance dependency, (O^{-16}). The cut-off radius of the potential is larger than the interatomic distance within the diamond structure of the indenter tip, therefore, no substrate atoms can penetrate into the indenter during the nanoindentation process. Between 1.8 Å and 1.6 Å, the value of the potential energy changes slightly, allowing an incremental propagation of the indenter by steps with the size of 0.2 Å.

Below 1.6 Å the potential energy increases dramatically, in this range the potential shows hard-sphere behaviour, simplifying it to a pure contact interaction and minimizing the increase of the system energy due to the presence of an external potential:

4.6.2 Metallic bond: Embedded Atom Method

Embedded **A**tom **M**ethod or **EAM**-potentials are used for the description of metallic material systems. The EAM potential was not used in this work, but is based on some interesting ideas for which it should be briefly described at this point.

The Embedded Atom Method was first proposed by Daw and Baskes in the mid 1980's in order to overcome the problems which appeared while using pair potentials for the description of metals. In detail, the usage of Lennard-Jones pair potentials was limited to a small group of noble gases, where no valence electrons are available and the interaction is based solely on the temporary asymmetric distribution of the electron density. This assumption leads to incorrect calculations of the ratio of the cohesive energy and melting temperature $E_c/k_B T_m$ (30 in metals, 10 in two-body systems), the ratio of the vacancy formation energy and cohesive energy E_v/E_c (0.25-0.33 in metals, about 1 in two-body systems) and of the ratio of two elastic constants C_{12}/C_{44} (1 in two-body systems, deviation from 1 in metals). The way out was the further development of the simple potential form of the Lennard-Jones potential, presented in equation 4.30, which was introduced in [106] and [138].

Equation 4.35 presents the form of the total energy of the system. The first term $\phi_{ij}(r_{ij})$ of equation 4.35 is the pair part or core-core repulsive term, already incorporated in the old model. The additional term $F_i(\overline{\rho_i})$ is the embedding or cohesive term representing the energy of the ionic core when inserted in the local electron density ρ_i. Since the metallic bond consists of delocalized electrons, the average electron density $\overline{\rho_i}$ is constructed as a superposition of the individual electron densities:

Section 4.6. Theory of interatomic potentials

$$E_{\text{tot}} = \frac{1}{2}\sum_{i,j} \phi_{ij}(r_{ij}) + \sum_i F_i(\overline{\rho_i}) \qquad (4.35)$$

$$\overline{\rho_i} = \sum_{i \neq j} \rho_{ij}(r_{ij}) \qquad (4.36)$$

The total energy of a metallic system E_{tot} is therefore defined as a sum of individual core-core interactions $\phi_{ij}(r_{ij})$ inserted into the electronic density $\overline{\rho_i}$. This new approach enabled a realistic prediction of many material properties of metals and is nowadays a standard tool for the description of metallic bonds for the use in molecular dynamics.

4.6.3 Covalent bond

It is not surprising that the development of the covalent potential functions went hand in hand with the increasing technological importance of semiconductors such as silicon and germanium. During the 80-s of the last century, many research groups devoted themselves to the development of suitable potential functions for these materials. All of the potential functions are multi-body potentials, a further development of simple pair potentials which is necessary to represent the angular dependency of covalent bonds. These potentials can be subdivided into two major categories: on the one hand those potentials, which can be split into two- and three-body terms and on the other hand, the bond-order potentials. The most prominent potential of the first class is the Stillinger-Weber potential [48], for the second class the best known representative is the Tersoff potential [74, 75, 76, 77, 78]. Both potentials consist explicitly of two pair potentials, one repulsive and one attractive hereby having a potential minimum corresponding to bond length. Covalent potentials are short distance functions, in their form they incorporate a cut-off function which limits their influence onto the nearest neighborhood. In sum, the attractive part of the Tersoff potential is multiplied by an enviroment-dependent parameter or a bond-order-parameter b_{ij}.

The origins of the bond-order parameter were set by Abel [49] using quantum mechanics and state that the more neighbors one atom possesses the weaker is the chemical bond to each of them. This statement holds for both metallic and covalent systems, the development of the modified embedded-atom potential for silicon is one of proofs of the same origins of these potential forms. The bond-order parameter makes it possible to describe different polymorphs of the same element, e.g., carbon in its sp^3 bonded form as diamond and sp^2 bonded form as graphite. Not only the two separate forms, but also the phase transition between two stabile configurations can be modelled by bond-order potentials making them an ideal tool for the description of systems far from thermodynamic equilibrium.

Tersoff potential

The bond-order Tersoff potential is one of the most widely used potentials for the simulation of covalent materials. Due to its simple parametric form it can easily be published for different materials, the implementation of the Tersoff potential can be found in almost every MD code. One of the first Tersoff potentials was published in [75] where the original author reported the simulation of different structural and energetic properties of carbon, such as diamond, graphite, quenched and liquid amorphous carbon. The author put special emphasis onto the investigation of elastic properties, phonons, defects and migration energies. Probably the most interesting feature of the Tersoff potential for carbon is its capability to describe different polytypes of carbon having only small structural energy differences such as diamond and graphite, see **Figure 4.3**.

For a system of interacting particles, here labeled with i and j the overall energy E corresponds to the sum of the individual energies E_i of atoms inserted in the potential field V_{ij} of all other atoms:

$$E = \sum_i E_i = \frac{1}{2} \sum_{i \neq j} V_{ij} \qquad (4.37)$$

The pair-like interaction V_{ij} can be further decomposed into the at-

Section 4.6. Theory of interatomic potentials

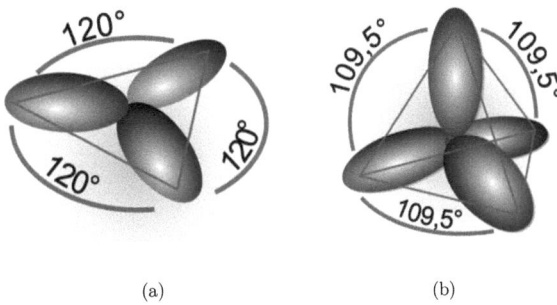

Figure 4.3: Schematic representation of planar sp^2 hybridization state in (a) as it can be found in graphite and the sp^3 hybridization state of the diamond tetrahedra in (b), source [4].

tractive pair term f_A, the repulsive pair term f_R and the cut-off function $f_c(r_{ij})$:

$$V_{ij} = f_c(r_{ij})\left[f_R(r_{ij}) + b_{ij}f_A(r_{ij})\right] \quad (4.38)$$

Both the attractive and the repulsive term of equation 4.38 are formulated as Morse functions [122] related to the exponential dependence of the electronic density:

$$f_R(r_{ij}) = A_{ij}e^{-\lambda_{ij}r_{ij}} \quad (4.39)$$

$$f_A(r_{ij}) = -B_{ij}e^{-\mu_{ij}r_{ij}} \quad (4.40)$$

Parameter B_{ij} describes the attractive force or the bond strenght caused by the formation of electronic pairs, while the repulsion, originating from the overlapping of inner electronic shells of individual atoms, is represented trough the parameter A_{ij}. Parameters λ_{ij} and μ_{ij} describe the exponential decay of both the attractive and repulsive force between particles i, j with the distance vector r_{ij}. In general, bond formation requires longer range of the attractive pair part $f_A(r_{ij})$ or $\lambda_{ij} > \mu_{ij}$.

Chapter 4. Molecular Dynamics (MD) method

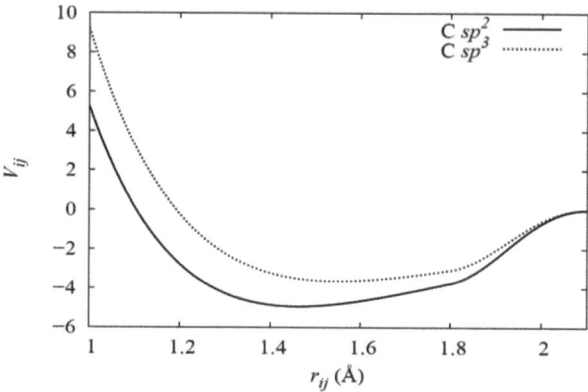

Figure 4.4: Tersoff pair potential part for two hybridization states of carbon. Notice the weaker bond in the case of the sp^3 state caused by the bond-order parameter due to the increased number of neighbours, source [108].

The very idea of the bond-order potentials is represented in the local-enviroment-dependent parameter b_{ij} which is multiplied by the attractive pair potential term in the equation 4.38. The bond-order parameter b_{ij} is responsible for the weakening or strengthening of the atomic bonds in dependence of the local enviroment. The b_{ij} is an empirical function and depends parametrically on:

- The distance r_{ik}, which is the distance between the atom i and its neighbour k. The index ij in the bond order parameter b_{ij} implies that the neighbor k will affect the bond between the atoms i and j;

- The angle θ_{kij} between the atom triplet i, j and k with the atom i in the middle;

- The chemical type of the atom k.

The exact definition of the bond-order parameter b_{ij} is given in following equations:

Section 4.6. Theory of interatomic potentials

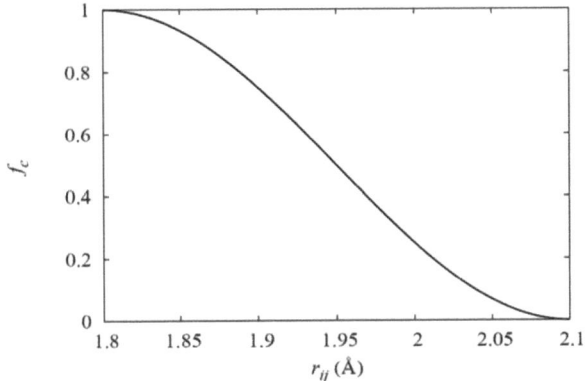

Figure 4.5: Cut-off function of the Tersoff potential for carbon, source [108].

$$b_{ij} = \chi_{ij}\left(1 + \gamma_i^{n_i}\zeta_{ij}^{n_i}\right)^{-\frac{1}{2n_i}} \qquad (4.41)$$

$$\zeta_{ij} = \sum_{k \neq i,j} f_c(r_{ik})g(\theta_{kij})\exp[\mu_{ij}^3(r_{ij}-r_{jk})^3] \qquad (4.42)$$

$$g(\theta_{kij}) = 1 + \frac{c_i^2}{d_i^2} - \frac{c_i^2}{d_i^2 + (h_i - \cos\theta_{kij})^2} \qquad (4.43)$$

The influence of the bond-order parameter is graphically represented in **Figure 4.4**, where the pair interaction of the Tersoff potential for carbon is depicted, in one case for the sp^3 hybridization state and the four-fold coordination as in the case of diamond and in the other case of an sp^2 hybridization state and the three-fold coordination as in the graphite structure.

The formation of electronic pairs and therefore of the covalent bond is highly distance dependent. Due to this, a cut-off function f_c is introduced into the Tersoff potential. Both the attractive and the repulsive

Chapter 4. Molecular Dynamics (MD) method

part of the interaction are multiplied by f_c restricting their influence range on a sphere of a certain radius around an individual atom, the contribution of the atoms outside of the cut-off radius is neglected. This greatly saves computation time since the number of pair interaction scales as $N!$ corresponding to the number of particles N.

The mathematical formulation of the cut-off function f_c can be parametrized as:

$$f_c(r_{ij}) = \begin{cases} 1 & r_{ij} < R_{ij} \\ \frac{1}{2} + \frac{1}{2}\cos\left(\frac{\pi(r_{ij}-R_{ij})}{S_{ij}-R_{ij}}\right) & R_{ij} < r_{ij} < S_{ij} \\ 0 & r_{ij} > S_{ij} \end{cases} \qquad (4.44)$$

For carbon, the schematical representation of the cut-off function is shown in **Figure 4.5**. It can be observed that the cut-off function f_c decreases from 1 to 0 in a short range from the atom.

In 1986 J. Tersoff reported another parameter set of the same functional form for the simulation of silicon [76], which is homopolar and isoelectronic to carbon. Potential parameters were improved several times by the original author in [74, 77] in order to represent better a wider range of material properties of silicon. In 1989 Tersoff published potential parameters for germanium. In the same paper he published the principle to calculate the potential parameters for multicomponent systems out of the parameters for single elements. In general, potential parameters for compound systems are arithmetic and geometric averages of the individual single element parameters. In the same paper, this principle was applied to SiC and SiGe binary systems and the material properties of SiC were analysed. This simple approach motivated further research groups into developing potential parameters for other materials and combining them with the existing ones in order to simulate material properties of new binary compounds. The research group around de Brito Mota developed a Tersoff potential for nitrogen [45] and applied it to Si_3N_4.

Not only binary systems, but also ternary systems were modeled using

Section 4.6. Theory of interatomic potentials

Tersoff potential. Matsunaga et al. report in [89] about the development of the Tersoff potential parameters for boron and simulation of cubic boron carbonitrides while using the previously developed Tersoff potential parameters for carbon and nitrogen. Next to boron carbonitrides, as introduced in [89], Si-C-N was another interesting material system modeled by the Tersoff potential. Resta reports in her PhD-Thesis [108] and in [109] about the molecular dynamics simulation of amorphous Si-C-N ceramics. Special emphasis was put on the thermodynamic properties of the material systems such as the phase separation into SiC, Si_3N_4 and amorphous carbon (a-C) under certain conditions.

Another ternary material system modeled by the Tersoff potential was Si-B-N. Griebel et al. report in [96] the simulation of BN-nanotubes in an amorphous Si-B-N-matrix. They used the original potential parameters for silicon reported by Tersoff in [74, 76, 77], as well as the potential parameters for nitrogen [45] and boron [89]. A summary of Tersoff potential parameters for silicon, carbon, nitrogen and boron is given in the **Table 4.1**.

Element/Parameter	Silicon	Carbon	Nitrogen	Boron
R (Å)	3.0	2.1	2.1	2.1
S (Å)	2.7	1.8	1.8	1.8
A (eV)	1830.8	1393.6	6368.14	223.01
B (eV)	471.18	346.7	511.76	171.29
λ (Å^{-1})	2.4799	3.4879	5.4367	1.72790
μ (Å^{-1})	1.7322	2.2119	2.7	1.38190
γ	$1.1 \cdot 10^{-6}$	$1.57 \cdot 10^{-7}$	$5.29 \cdot 10^{-3}$	$1.60 \cdot 10^{-6}$
n	0.78734	0.72751	1.33041	3.99290
c	100390	38049	20312.0	0.52629
d	16.217	4.384	25.5103	$1.587 \cdot 10^{-3}$
h	-0.59825	-0.57058	-0.56239	-0.5

Table 4.1: Tersoff potential parameters for the calculated quaternary Si-C-N-B system.

Recently, Munetoh et. al [132] used the *ab initio* method to develop Tersoff potential parameters for oxygen and applied them with success to the simulation of SiO_2. Instead of representing the long range Coulomb interaction of the oxygen atom, the authors incorporated the polarisability of the oxygen atom in the local force calculation. Despite this simplification, the potential was able to describe different polymorphs of SiO_2 as well as individual phase transformations.

4.6.4 Bond ionicity

Single crystal carbon, silicon and germanium consist of one atom type, therefore, chemical bonds between individual atoms are purely covalent, the charge distribution along the connection line is symmetric and the bond is nonpolar as for H_2 and Cl_2 as shown in **Figure 4.6(a)**. The purely covalent nature of the chemical bond in the case of one component systems allows the description of the chemical bond within the frame of the Tersoff potential.

Compound ceramics such as SiC, Si_3N_4 or c-BN consist of different atom types, thus individual atom types having different electronegativity. The electronegativity or the affinity of an atom to bind electrons is smallest at the left of the periodic table, where atoms have only one weakly bonded electron in addition to completed inner electronic shells. This additional electron is released relatively easy, elements such as Li, Na, K, Ca always appear in the form of their positively charged cations Li^+, Na^+, K^+ and Ca^{2+}. The electronegativity increases to the right of the periodic table and is largest for halogens, which attract electrons stronger in order to complete their electronic shell. Electronegativity decreases going from top to the bottom of the periodic table, which is reasonable considering the larger number of shells and looser bond of outer, valence electrons.

In the extreme case of a bond of two atoms of highly different values of electronegativity, an ionic bond is a result, as in the case of a rock-salt structured NaCl. In most cases, the chemical bond of two elements of

Section 4.6. Theory of interatomic potentials

different electronegativity is a mixture of a covalent and an ionic bond, the electron transfer in such bond is not complete, the bond is asymmetrical regarding the electronic distribution. In the case of SiC and Si_3N_4 such a situation is present (see **Figure 4.6(b)**), this makes it necessary to quantify the amount of ionic character in an individual chemical bond A-B.

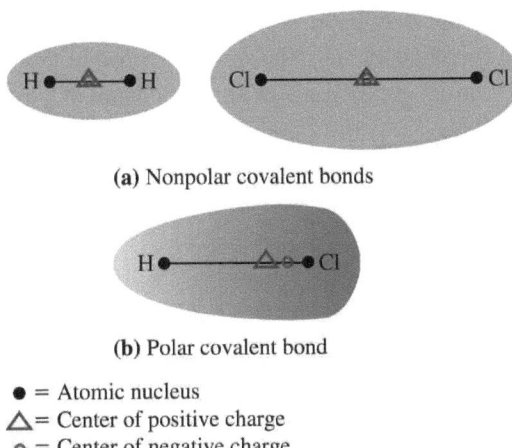

(a) Nonpolar covalent bonds

(b) Polar covalent bond

- ● = Atomic nucleus
- △ = Center of positive charge
- ○ = Center of negative charge

Figure 4.6: Representation of the electronic density for nonpolar covalent bonds (H_2, Cl_2) and a polar covalent bond (HCl).

Three methods are currently used in order to do this, Pauling, Phillips and Harrissons ionicity scale. These methods have some conceptual differences, which makes the comparison of ionicity values accross the scales difficult, however, more interesting is the comparison of ionicities of two different material systems within one ionicity scale.

Phillips studied the chemical bonding properties of $A^N B^{8-N}$ crystal structures and their electronic energy-band states. His ionicity scale f_i is defined in terms of average quantities such as the homopolar E_h and heteropolar parts C of the complex energy gap E_g associated with the A-B bond in the crystal [128]:

Chapter 4. Molecular Dynamics (MD) method

$$E_g = E_h + iC \qquad (4.45)$$

In this formulation, the Phillips bond ionicity f_i is then defined as:

$$f_i = \frac{C^2}{E_g^2} = \frac{C^2}{E_h^2 + C^2} \qquad (4.46)$$

Pauling based his definition of ionicity scale on empirical heats of formation. He formulated that the difference in electronegativity (X_A and X_B) of individual compounds in an A-B chemical bond is the source of the ionicity. Therefore, the bond ionicity defined by Pauling f_i^P is a function of the individual electronegativities X_i-s:

$$f_i^P = 1 - \exp\left(-\frac{(X_A - X_B)^2}{4}\right) \qquad (4.47)$$

Paulings scale ranges from zero for homopolar bonds with no ionicity, e.g. H-H to one equaling the ionic bond, where a complete charge transfer from one atom to its bond-partner has occured, e.g. NaCl.

The ionicity scale defined by Harrisson f_i^H uses the energy gap between the bonding and antibonding states (similar to the definition of Phillips) as well as the energy change for transfering an electron from anion to cation for its definition:

$$f_i^H = \frac{V_3}{\sqrt{V_2^2 + V_3^2}} \qquad (4.48)$$

with V_2 is the half of the splitting between bonding and antibonding states and V_3 is the half of the energy change in transferring an electron from anion to cation.

The summary and comparison of three stated ionicity scales for semiconductors of group IV, III-V and II-VI is made by Adachi in [128], while a short overview is given in **Table 4.2**. It is clear that SiC shows a very low ionicity on all three scales, which makes the approximation of modeling only the covalent character of the Si-C bond justified. The analysis of the ionicity of Si_3N_4 is a bit more difficult, since the sum of

Section 4.6. Theory of interatomic potentials

stoichiometry numbers is 7 rather than 8, as in Paulings definition.

In [22] the authors presented a first-principles pseudopotential study of the structural and electronic properties of β-Si$_3$N$_4$ as well as of the hypothetical compound β-C$_3$N$_4$. **Figure 4.7** presents the normalized charge density distribution along the C-C, C-N, Si-C and Si-N chemical bonds.

System	Material	f_i	f_i^P	f_i^H	System	Material	f_i	f_i^P	f_i^H
IV	Diamond	0	0	0	II-VI	MgO	0.841	0.88	
	Si	0	0	0		MgS	0.786		
	Ge	0	0	0		MgSe	0.790		
	Sn	0	0	0		MgTe	0.554		
	SiC	0.177	0.11	0.35		ZnO	0.616	0.80	0.69
						ZnS	0.623	0.59	0.69
III-V	BN	0.221	0.42	0.43	II-VI	ZnSe	0.630	0.57	0.70
	BP	0.032				ZnTe	0.609	0.53	0.68
	BAs	0.044				CdS	0.685	0.59	0.74
	AlN	0.449	0.56	0.57		CdSe	0.699	0.58	0.74
	AlP	0.307	0.25	0.47		CdTe	0.717	0.52	0.76
	AlAs	0.274	0.27	0.44		HgS	0.790		
	AlSb	0.250	0.26	0.56		HgSe	0.680		
	GaN	0.500	0.55	0.61		HgTe	0.650		0.78
	GaP	0.327	0.27	0.48					
	GaAs	0.310	0.26	0.47					
	GaSb	0.261	0.26	0.43					
	InN	0.578							
	InP	0.421	0.26	0.55					
	InAs	0.357	0.26	0.51					
	InSb	0.321	0.25	0.48					

Table 4.2: Phillips (f_i), Pauling (f_i^P) and Harrison ionicities (f_i^H) for number of group-IV, III-V and II-VI-semiconductors, source [128].

Chapter 4. Molecular Dynamics (MD) method

In case of C-C the charge distribution is symmetric as expected, while a small asymmtery is present in the case of Si-C and C-N. Si-N bond shows significantly higher asymmetry of the charge density distribution, implying that the modelling of Si-N bond by a pure Tersoff covalent potential may overextend the approximation.

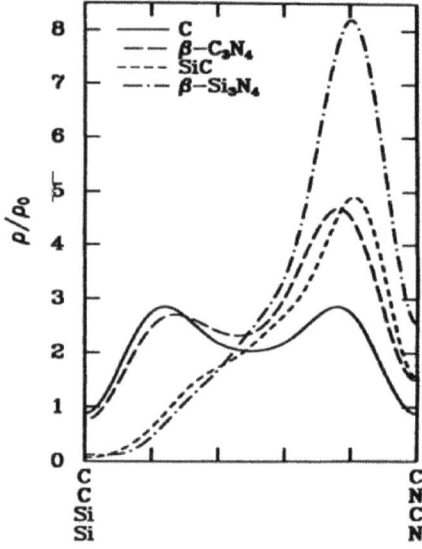

Figure 4.7: Total valence charge density along the chemical bond in diamond, β-C_3N_4, SiC, and β-Si_3N_4, normalized by the number of valence electrons per unit cell, ρ_0. All bond lengths are normalized to the same length, source [22].

In [88] Albe is confronted with the same problem, while developing an interatomic potential for the BN material system. The chosen form of the potential presents only the covalent nature of the B-N bond, while the ionicity is neglected. **Table 4.2** makes it evident that the ionicity of the B-N bond is in all cases higher than in the case of the Si-C bond, which motivates the intention to consider only the covalent

Section 4.6. Theory of interatomic potentials

nature of the chemical bond for the SiC potential. At this stage, a few arguments of Albe for the choice of the potential form are presented, same assumptions are also valid for the case of SiC and Si_3N_4:

1. SiC zincblende structure is similar to the diamond structure of both carbon and silicon, with the lattice constant being approximately the arithmetic average of the lattice constants of carbon and silicon. This argument corresponds also to the Pauling definition of ionicity, in SiC, both Si and C being four-fold coordinated, the covalent nature of the chemical bond is dominant.

2. The Phillips ionicity f_i of SiC is f_i=0.177 and charge distribution asymmetry coefficient g=0.475. A direct comparison of these values to values of BN (f_i=0.256 and g=0.484) for which the Tersoff potential is also used makes it evident that clear statement about the possibility to describe the Si-C bond within the concept of a bond-order potential.

3. Neither optical nor electrical properties of the material will be modelled.

4. An explicit incorporation of a Coulomb term in order to describe the asymmetrical charge distribution within a SiC bond would make the modeling very slow. Representation within a shell model prerequisites additional virtual particles representing a center of charge density along the chemical bond. This approach would significantly increase the computational time, while only a minor improval with respect to the representation by a simple covalent bond can be expected.

5 Simulation of the sputtering process

5.1 Introduction

Coating by SiC and Si_3N_4 requires the removal of individual Si and C or N atoms out of the target material for subsequent deposition onto the desired substrate. Several methods such as laser ablation [134, 135], chemical vapor deposition CVD via plasma etching [63, 114, 136, 153], and physical vapor deposition PVD [64, 79, 85, 90, 123, 150] can be used in order to achieve this objective. In this chapter a brief insight into the experimental and simulation aspects of the PVD magnetron sputtering process will be given.

A back sputtering process or shortly sputtering is the removal of target material by the bombardment by energetic ions. In the scientific community the energy range of the incident ions is normally chosen to be between 0.5 eV and 20 keV, hereby covering the operating energies of ion guns used for sputter cleaning, sputter depth profiling, sputter excitation, Auger electron spectroscopy AES , X-ray photoelectron spectroscopy XPS, secondary ion mass spectroscopy SIMS , sputter deposition, plasma deposition etc. [80]. Within this work, we concentrate onto the low energy regime between 20 eV - 1000 eV which is used for the subsequent sputter deposition process.

Sputtering can be divided in reactive and non-reactive sputtering. In the case of reactive sputtering the incident ion reacts chemically with the surface atoms of the target material, bonds within the target material are broken while new chemical bonds are formed, mostly containing the

Chapter 5. Simulation of the sputtering process

incident ion as reactant. This can be the case if incident ions are very reactive elements such as in the case of hydrogen-containing process gases, e.g., methane CH_4, silane SiH_4 or HMDS (hexamethyldisilazane [3]) $[(CH_3)_3Si]_2NH$. Reactive magnetron sputtering can, therefore, lead to a high content of deposited hydrogen, which can have an effect onto the structural stability of the coating [28].

In the case of non-reactive magnetron sputtering, the interaction between the incident ion and the target material is of purely physical nature: momentum and energy are transferred by collisions, chemical reactions between the interacting particles can be neglected. Depending on the reactivity of the target material and the incident ions, the sputtering process is more reactive or non-reactive. Due to their chemical inertia, noble gases, such as argon, are often used in experiment in order to prevent any chemical reaction during the sputtering process.

5.2 Fundamental processes

It is necessary to understand the underlying interaction between the incident ion and the target surface in order to describe a wide variety of scenarios which can result.

As the distance between the incident ion and target atoms decreases, electrons in outer shells of neighbouring atoms interact, individual atomic orbitals overlap to some extend and evolve into molecular orbitals of the *target atom/incident ion* short life molecule. Individual atomic orbitals are however energetically favored by the electrons of interacting atoms, especially in the case of noble gas atoms, therefore, resulting into a repulsive force between the individual atoms, known as the *electronic stopping power*. Since the noble gas atom is a positively charged ion, with high electronegativity, it captures an electron from the target atoms resulting into collisional ionization, see upper left part of **Figure 5.1**. Therefore, an initially positively charged ion can very well be scattered as a neutral atom.

If the kinetic energy of the incident ion is large enough, the electronic

Section 5.2. Fundamental processes

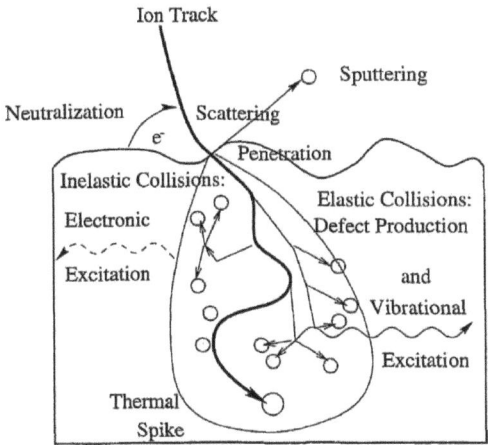

Figure 5.1: Processes occuring along the ion trajectory, source [88].

stopping power will be surmounted and the shortest distance between the incident ion and the target atom will decrease further. As the distance between individual atomic nuclei further decreases, molecular orbitals of the target atom-incident ion molecule will merge into atomic orbitals of an atom with the atomic number $Z_{target\ atom} + Z_{ion}$ and the electrons will be promoted further into higher energy states of the unified atom. In addition, positively charged atomic cores of the individual atoms will result into a repulsive force: this effect is known as the *nuclear stopping power*. Both described processes are fast and occur on a time scale of 10^{-13} s [80, 88].

Depending on the impact angle, surface impact coordinate and impact energy, the incident ion will evolve differently, it can either be reflected from the target surface, adsorbed by it or it can penetrate further into the target material. In the last case, the incident ion will loose its kinetic energy during elastic and inelastic collisions with target atoms. Possible results are electronic excitations and defect productions such as vacancies, dimers and back sputtered atoms. At some point the incident ion will loose all of its initial energy and remain trapped in the target bulk. The collision cascade, caused by the incident ion will propagate through the crystal, its spatial expansion is larger than the ion trajectory itself.

The kinetic energy of the incident ion is several scales larger than the average thermal energy of the target atoms, all of the mentioned processes occur far from thermodynamic equilibrium. The collision cascade evolves over several 100 fs, the energy deposition into a well defined area along the ion trajectory is very large and is often referred to as *thermal spike*. At a later stage in this work, a more refined definition of the term spike will be introduced. After the initial stage, the deposited energy will dissipate over the crystal lattice for several picoseconds, corresponding to the heat dissipation of 10^{14}-10^{15} K/s [88].

5.3 Sputtering mechanisms

There are 3 possible mechanisms which can lead to physical sputtering: the few-collisions or single knock-on regime, linear cascade regime and the spike regime. At the lower energy range, near to the sputter threshold energy the singe knock-on or few collisions regime is dominant. Under this regime the impacting ion directly transfers a large amount of kinetic energy to a few surface-near atoms, enabling them to overcome the binding forces of the target material. This directly leads to back sputtering. The single knock-on regime is schematically represented in **Figure 5.2 a**. The schematic makes clear, that only few layers beneath the target surface are involved in the collision, which is an indication of direct impacts between the collision partners. At higher ion energies, the kinetic energy of the incident ion suffices not only to cause single knock-out of a specific near-surface atom but also to cause a collision cascade between the atoms of the target material. This cascade will result in back sputtering if the cascade region overlaps with the surface of the target material and the kinetic energy of the surface atoms is higher than the surface binding energy U_0 as shown in **Figure 5.2 b**. The last

Section 5.3. Sputtering mechanisms

regime is the so called spike regime: the main difference between the collision cascade and the spike regime (shown in **Figure 5.2 c**) is the cascade density: in the case of a linear cascade only a certain number of atoms in a specific volume is in motion, while in the spike regime the majority of atoms is affected. The term spike has the same origin as in *thermal spike* but is used in a more general way at this stage: it reflects the idea of a large number of particles sharing the same characteristic, another example would be a *displacement spike*.

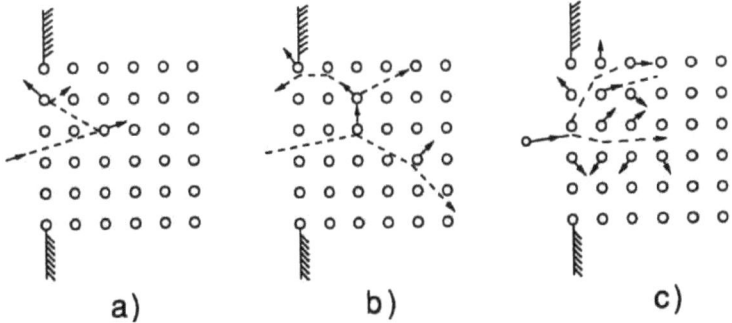

Figure 5.2: Three mechanisms of back sputtering by elastic collisions, source [80]. At the lower energy range, near the sputter threshold energy, in general around 50 eV, the single knock-on regime, presented in a) is dominant. In b) the linear cascade regime is presented, involving a large number of atoms with a random motion. The energy range of the linear cascade regime can go up to several keV-s for a single ion impact. At a very high energy and/or impact rate the spike regime, presented in c) is the dominant process. Within the spike regime, large number of atoms within a certain unit volume perform similar motion, e.g., in the same direction. The spike regime is the dominant back sputtering process at ion impact energies in the order of several keV-s.

Chapter 5. Simulation of the sputtering process

5.4 Sputter threshold energy and sputter yield

The concept of sputter threshold energy and of the sputter yield has to be introduced in order to quantify the sputtering process: the sputter threshold energy is the minimal kinetic energy of the incident ion needed to cause a sputtering event, while sputter yield is the ratio of the number of back sputtered particles of the target material and the number of incident ions. For most materials, the sputter threshold energy lies in the range of 5-40 eV and the sputter yield in between 10^{-5} and 10^3. Since the sputter yield can obtain values which are several orders of magnitude < 1, it is evident that back sputtering has a highly statistical nature and that not all ion impacts lead to back sputtering, even it the threshold energy is surmounted.

At the macroscopic scale and for high incident energies, an experiment is the most reliable method for evaluating the sputter yield; given the flux of incident ions and knowing the atomic mass of the target material the calculation is simple, the only remaining variable is the mass difference of the target before and after the experiment. However, experimental estimation of the sputter yield becomes unreliable at lower ion energies, near the sputter threshold energy where the sputter yield becomes very low ($< 10^{-3}$) and the measurement is unreliable due to limited accuracy of the experimental apparatus.

5.5 Sigmunds sputter theory

Next to the experimental measurement of the sputter yield, several theoretical attempts towards the correct estimation of the sputter yield for various materials have been undertaken over the years, one of the best known is Sigmunds sputter theory [116].

Sigmunds sputter theory assumes an amorphous or polycrystalline target material of infinite extent. The behavior of incident ions and recoil atoms can be described by the Boltzmann transport equation.

Section 5.5. Sigmunds sputter theory

Sputtering is caused by quasi-elastic collisions, while chemical reactions and inelastic collisions leading to possible ionization effects as described in **Section 5.2** of this chapter, **Figure 5.1**, are neglected. This assumption reflects the experimental situation in the case of chemically inert noble gas sputtering by far extent, making the validation of experimental results by the Sigmunds sputter theory very frequent in the literature.

The collision cascade caused by an incident ion is divided into two stages, at first stage the single knock-on effect dominates, where the ion transferes a large amount of its kinetic energy onto a few near surface target atoms and the second stage in which the kinetic energy dissipates over the large number of low energy recoil atoms. Due to this separation, the sputter yield can be very sensitive to the surface impact coordinate of the incident ion in the case of a low ion energy and open crystal structures of the target material, e.g., silicon. A special emphasis to this effect is given in **Chapter 7** where different sputter yields are observed for different impact coordinates. Since sputtering is only a collision-based event, the ion incident energy, the cross-section of the ion-target atom collision, target atom-target atom collision cross-section and the surface binding energy are the only input parameters for the estimation of the sputter yield.

Sigmunds sputter theory disregards the last stated sputter regime, namely the spike regime. This approximation is valid at lower impact energies and lower ion masses, but the effect of the thermal spike has to be taken into account in the case of high energy bombardment or bombardment by heavy ions. In order to overcome this limitation, Sigmund and Szymonski evaluated in 1984 the temperature dependence of the sputter yield of metals and insulators and have shown in [117] a dramatic increase of the sputter yield of silver at temperatures near the melting point at 1235 K caused by impacting ions, the reader is referred to **Figure 7** in [117].

Chapter 5. Simulation of the sputtering process

According to Sigmund, the sputter yield Y of an impacting-ion/target pair can be deduced from the equation:

$$Y = \frac{3.56}{U} \cdot \frac{Z_t Z_i}{\sqrt{(Z_t^{2/3} + Z_p^{2/3})}} \cdot \frac{M_i}{(M_t + M_i)} \cdot \alpha(M_t/M_i) S_n(E/E_{ti}) \quad (5.1)$$

with the reduced energy E_{ti}:

$$E_{ti} = 0.0308 \cdot (1 + M_i/M_t) Z_t Z_i \sqrt{Z_t^{2/3} + Z_i^{2/3}} \quad (5.2)$$

Z_t is the atomic number of the target material, Z_i is the atomic number of the incident ion, M_t and M_i are their atomic masses. E presents the ion impact energy and U the surface binding energy of the target material. α is a function of the ratio of the atomic masses M_t/M_i and of the incident angle but not of the incident energy. Function S_n is an universal function, known as the nuclear stopping cross section of the reduced energy E_{ti}.

Equation 5.1 contains only the atom masses (and numbers) of the incident ion and target atoms, implying Sigmunds initial intention to describe the sputtering behavior of single element materials. A generalization towards the description of the sputtering of compound materials is done in rather simple manner and has its limitations: the *target atom mass* is taken as the arithmetic average of individual compounds.

Nevertheless, as it will be shown in the following sections, Sigmunds sputter theory describes quite well the sputtering behavior of a large number of materials.

From Sigmunds sputter theory, equation 5.1, several important statements can be derived. It is evident that the sputter yield is proportional to the ratio of the atomic masses of the incident ion and the target material M_i/M_t.

This would lead to 3 immediate assumptions for all remaining parameters fixed:

Section 5.5. Sigmunds sputter theory

1. Bombardment by heavier ions causes larger sputter yield.

2. Materials with larger atomic masses have a smaller sputter yield than those with smaller atomic masses.

3. For compound materials, e.g., SiC: the lighter compound is more probable to be back sputtered, the differentiated sputter yield of individual compounds will not necessarily reflect the stoichiometry of the target material.

The first assumption can be easily validated by the literature results. Ecke et al. calculated in [50] the sputter yield of silicon and silicon carbide caused by Ne^+, Ar^+ and Xe^+ ions (atom masses 20.2, 39.9 and 131.3) under a 60 ° incident angle to the surface normal. Ion incident energies were set to 0.5 keV, 1 keV and 5 keV. Both Si and SiC showed higher sputter yield under bombardment by heavy Xe^+ than by other two ions. For an exact comparison one is referred to **Table 1** in [50]. Calculations were based on TRIM simulation package [86, 155] and were verified by experimental results. The impact of lighter ions such as H^+, $deuterium^+$ and He^+ is especially studied in the framework of nuclear technology because of the damage caused to the walls of reactor chambers and is presented in [147].

In the case of a bombardment by He^+, the material responded by a higher sputter yield then it was the case for H^+ and $deuterium^+$. An increased back sputter yield in the case of $deuterium^+$ bombardment in comparison to single proton or H^+ bombardment, reveals the sensitivity of the sputter yield onto the mass of the incident ion. Therefore, the initial statement also holds for the case of lighter ions, ensuring the completeness of Sigmunds theory for the whole range of noble gas ions used for non-reactive sputtering.

For the validation of the second statement we again turn to [50] as well as to present simulation results. Although silicon has lower surface binding energy than silicon carbide (4.7 eV vs. 6 eV) it shows smaller sputter yield, see again **Table 1** in [50], it is, therefore, evident that this

Chapter 5. Simulation of the sputtering process

effect is based solely on the atomic mass of silicon. This is also verified by present simulation results of bombardment of (100), (110) and (111) crystal surfaces of Si and β-SiC by argon ions (see **Figures 7.5** and **7.12 - 7.21** in **Chapter 7**).

Preferential sputtering is often stated as a reason for the formation of non-stoichiometric coatings or coating components from an originally single crystal target material of ideal stoichiometry. Sigmunds theory enables the interpretation of the second assumption once more, this time instead of comparing sputter yields of two different single element target materials, it can be used to describe the sputtering behavior of a compound material, where individual compounds have different atomic masses. In the case of β-SiC both Si and C atoms have the same surface binding energy of $E_{bind} = 6$ eV and since there are only Si-C bonds, it is only the atomic mass that can, according to our interpretation of Sigmunds sputter theory lead to preferential sputtering of carbon. Indeed, both carbon- and silicon-terminated SiC (100) crystals showed higher sputter yield of carbon as presented in **Chapter 7**, **Figures 7.12 - 7.21** and in [19]. The same statement holds for sputtering of Si_3N_4 by argon ions where the sputter yield of nitrogen is more than 4/3 times larger than of silicon, see again **Chapter 7**, **Figures 7.25-7.30** and research paper [19].

Figure 5.3 represents the sputter yield of five different hypothetical target materials with specific atomic numbers Z_1, normalized to the surface binding energy of 1 eV. It can be observed that for all atomic masses of target materials, the sputter yield starts at zero value at the sputter threshold energy and increases over 2-3 orders of magnitude in value until it reaches its maximum between 10 keV and 1 MeV. At higher energies, the curve decreases rapidly. Both the maximum sputter yield and its position on the x-axis are material dependent. The sputter yield of a real material can be obtained dividing it by its surface binding energy. The position of the maximum on the x-axis changes in the manner that materials with larger atom numbers have a maximum at a higher ion incident energy and vice versa. **Figure 5.3** shows only in a

Section 5.5. Sigmunds sputter theory

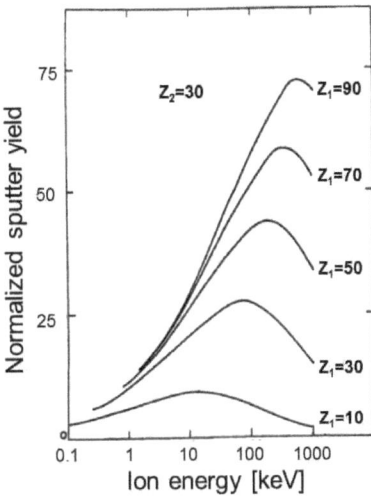

Figure 5.3: Sputter yield as a function of ion impact energy with ion atomic mass $Z_2=30$ (argon atom mass equaling 39.948) and different atomic masses of the target material normalized to the surface binding energy of 1 eV, source [80]. For compound semiconductors the atomic mass of the target material is an arithmetic average of individual compounds.

schematic way the principal behavior of the sputter yield and should be taken with caution.

As already stated, the sputtering process is governed by the single knock-on mechanism at the lower energy range near the sputter threshold energy. As the energy increases so does the collision probability and the collision cascade or even spike regime become the dominant mechanism in the sputtering process. However, the penetration depth of the incident ion also increases with the ion incident energy. This leads to an increased deposition of ions initial kinetic energy deeper into the crystal lattice of the target material. However, an overlapping of the collision cascade with the target surface is needed for a sputter event to occur, this becomes less probable at some stage, since the collision

Chapter 5. Simulation of the sputtering process

cascade is located deeper in the crystal bulk. A very good validation of the nature of this process will be shown later in this work on several examples where the argon impact coordinate overlaps with a free space on the crystals surface, so that single knock on events can be excluded. These cases show higher penetration depths of argon ions (see **Figures 7.4** and **7.9**) and a sputter yield which is practically zero for all impact energies, **Figures 7.2** and **7.8** in **Chapter 7**.

Later comparisons with present sputtering data will show that no maximum was obtained within the energy range under investigation (20-1000 eV). It is clear from **Figure 5.3** that this maximum could have been observed in the best case of c-BN (average atom mass 11.5) sputtering at ion energy of 8 keV. For the sputtering of heavier targets, such as Si (atom mass 28), SiC (average atom mass 20) and Si_3N_4 (average atom mass 20), the maximum of the sputter yield should be expected in the range of 10-100 keV. Considering the experimental situation, where 1 keV was the maximum incident energy, the behavior of the sputter yield over a wider energy range was not within the scope of this work.

5.6 Simulation of the sputtering process

Aside from experimental methods and theoretical estimations, simulation methods for the numerical simulation of the sputtering process evolved with increasing computing power. One of the first program codes available was kinetic Monte Carlo code, based on binary collision approximation (BCA), developed in the late 1970-s by J. P. Biersack in order to calculate the transport of energetic ions in amorphous targets [86]. Ziegler contributed over the next years to further program development, introducing together with the original author and U. Littmark a pair potential for the interaction of energetic noble gas ions with target atoms, see **Section 4.6.1**. The potential is nowdays known as the Ziegler-Biersack-Littmark(ZBL)-potential and the computer code **TR**ansport of **I**ons in **M**atter (**TRIM**). Both are most widely cited in the literature for the calculation of sputter yields for different materials.

5.7 Validation and extension of the sputtering results

Reliable validation of theoretical predictions and simulation data by experimental results is a crucial link needed to justify the existence of the simulation as a research method, while in return, simulation data is needed for a correct interpretation of unexpected experimental results. The probably best example is the channeling effect: the impacting ions penetrate deeper into the crystal structure due to the formation of long channels parallel to the trajectory of the impacting ion. This results into a sputter yield which is much smaller than a theoretical prediction for the corresponding impact ion energy.

In the case of the sputtering process, the simulation by MD/MC and experimental results are different by several orders in magnitude in time and length scales. This has severe impacts onto how results between both approaches should be compared. In an experiment, polycrystalline materials having different grain sizes and orientations are present. On the MC/MD scale however, one uses perfect single crystals with well defined crystallographic symmetries. If the trajectory of an incident ion beam is parallel to one of the more densely packed planes of atoms, the sputter yield will be far below experimental values, while it can reach values which are much higher than those measured in experiment if the ion trajectory is perpendicular to densely packed planes [107].

In the case of Sigmunds sputter theory, equation 5.1 and 5.2, the factor α can also incorporate angular dependency. Sigmunds sputter theory, however, lacks the information of the crystal surface orientation and the sputtering data obtained in this way is a homogenization of all possible crystal surfaces and all impact coordinates onto an individual crystal surface. This requires an evaluation of the back sputter yield by molecular dynamics simulations for different crystal orientations and surface impact sites.

Next to the angular and crystal orientation dependency of the sputter yield, a question should be allowed up to which extend amorphisation

Chapter 5. Simulation of the sputtering process

effects play a role in the sputtering process. Sputter experiments usually require several hours of measurement, depending on the ion energy, irradiation effects and amorphisation of the target material will be of a certain importance. Taking the ion rate, ion energy and target temperature, the right answer becomes highly parameter-dependent. Although the initial target my be a perfect single crystal, a constant bombardment by high energy ions will to an amorphisation of the target surface. It is not clear if an amorphous and a crystalline surface show the same sputtering behavior, and if not, how fast do amorphisation effects influence the sputter yield. Not taking amorphisation effect into account would lead to a sputter yield changing during the experiment, although all parameters remained unchanged.

In [52] Wehner investigated sputtering of metals by low-energy (up to 300 eV) Hg^+ ions. Special emphasis was given to the sputter yield dependency onto the crystal orientation of the target material and the distinction between the crystalline and amorphous state. This experimental investigation was done in a rather simple, but profound manner: a disc placed in front of the target surface collected all back sputtered particles. An amorphous target material would result into a diffuse, radially symmetrical pattern on the disc, while a more structured pattern would originate from a single crystal target material. Observations were similar to those made for silicon, presented in [55], see **Figure 5.4**. In the case of a single crystal material, the preferential direction of the ejection of sputtered particles is that of the closest packing, e.g [110] in fcc and [111] in bcc metals. In addition, it could be shown that the sputter threshold energy is considerably lower for an oblique than for a normal incident angle. **Table 1** in [52] gives an overview of sputter threshold energies for normal incident angle for different metalic targets and the whole range of noble gas incident ions (from Ne^+ to Xe^+) and Hg^+, as an example, 27 eV sputter threshold energy for a silicon bombarded by Ar^+ ions should be stated, which is supported by present simulations, see **Chapter 7**.

In [56] Anderson, Wehner and Olin discussed the temperature depen-

Section 5.7. Validation and extension of the sputtering results

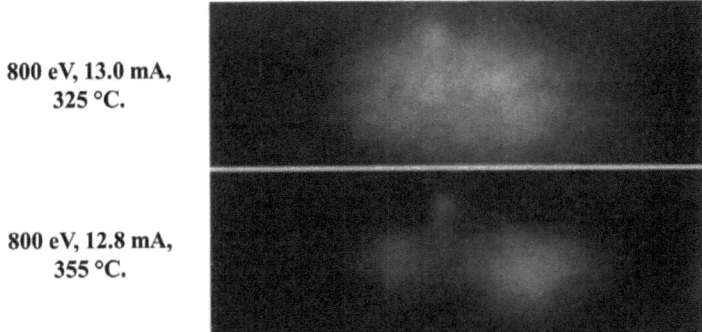

800 eV, 13.0 mA, 325 °C.

800 eV, 12.8 mA, 355 °C.

Figure 5.4: Typical atom ejection patterns in the ($1\bar{1}1$) plane for Ne$^+$ ion bombardment of the Si (111) surface, source [55].

dence of sputtered particle trajectories arising from Ne$^+$, Ar$^+$, Kr$^+$, Xe$^+$ bombardment of single crystal Ge targets. The energy range of impacting ions was chosen to 100-800 eV while target temperatures were in the range 0 °C - 350 °C. Specific ejection patterns remaining onto a collector placed in front of the target could be observed only above a specific temperature T_a, see **Figure 1** in [56]. These patterns were associated with the presence of a crystal structure in the uppermost target layers, while the absence of specific ejection patterns was interpreted as sputtering of an amorphised structure. Since the initial target was a Ge single crystal and the experiment duration was several hours, there was only one logical interpretation of these results: damage caused by impacting noble gas ions and annealing of the crystal structure in the uppermost layers of the target material are two competing processes, the first one being driven by the ion rate and impact energy and second the one driven by the target temperature.

In a subsequent publication of Anderson and Wehner [55], the authors extended the material system to Si, InSb and InAs (initially only Ge was investigated) and slightly the target temperature range to

Chapter 5. Simulation of the sputtering process

0 °C - 500 °C (up to 350 °C previously). As the material system evolved, so have also the results: In the case of silicon, the annealing temperature was determined to be 700 K for Ar^+ ion impacts and 620 K for the lighter Ne^+ noble gas. In addition, the critical ion impact energy E_c for Si was determined to be 400 eV. This energy is far above the sputter threshold energy of Si and presents the critical energy below which the damage caused by impacting ions is located only near the target surface, above this energy, a temperature higher than 700 K is required to heal the damage caused to the Si bulk material. **Figure 5.4** presents the ejection patterns in the $(1\bar{1}1)$ plane for Ne^+ ion bombardment of the Si (111) surface. Depicted are results for two temperatures, one below and one above the annealing temperature T_a=620 K (347 °C), as presented, the second pattern of back sputtered particles showes some structure, reflecting the crystal symmetry, while the ejection pattern of the amorphous silicon structure is diffuse.

In [53] Anderson further pursued the original idea of Wehner. The material system of Si, Ge, InSb and InAs was further extended to include GaAs and CdTe. For GaAs the critical annealing temperature for Ar^+ and Ne^+ bombardment was determined to be around 130 °C. An equivalent for CdTe could not be found, although the temperature range went up to 510 °C. At this temperature a blue glow in the plasma was observed indicating a Cd evaporation from the target material. In this publication the author also withdraws the initial concept of a critical ion energy E_c, they state E_c to be an erroneous conclusion resulting from the method used. In his work the author describes bombardment of GaAs by 500 eV Ar^+ ions above the critical temperature of 130 °C in order to preserve the crystal structure and investigate possible orientation effects. In a diamond structure, the <111> direction is most densely packed, an increased sputter yield is therefore expected for this crystal orientation. In the case of GaAs, having diamond like zincblende structure, one should discriminate between two different most densely packed directions, namely the <111> with the stacking order GaAsGaAs and Ga termination and the $<\bar{1}\bar{1}\bar{1}>$ AsGaAsGa with As termination. The

Section 5.7. Validation and extension of the sputtering results

<$\overline{111}$> direction resulted in approx. 10 % more ejected material, the analysis of the ejected material for both surfaces showed that the element terminating individual crystal orientation (top most layer) is more likely sputtered (ratio approx. 55 %-45 %), indicating that sputtering is surface dominated process.

A contrary result was obtained 1973 by Cooper et al. in [29] where an experimental analysis of the <111> and <$\overline{111}$> faces of GaAs (differing only in surface termination) by 100 eV Ar$^+$ ions was performed. The author reported a 0.169 molecules/ion sputter yield for the <111> surface and approx. 6 % smaller sputter yield for the <$\overline{111}$> surface. Comas and Cooper in [68] performed an experimental study of the sputtering process of (110), (111) and ($\overline{111}$) GaAs faces by 0-140 eV Ar$^+$ ions. The analysis of sputter yields Y showed Y(111) = Y($\overline{111}$) > Y(110). The works of Comas, Cooper, Anderson and Wehner were the first dealing with the crystal orientation dependency of the sputter yield. The same observation was made in the framework of the present research on the example of Si and β-SiC which possess the same diamond like crystal structure.

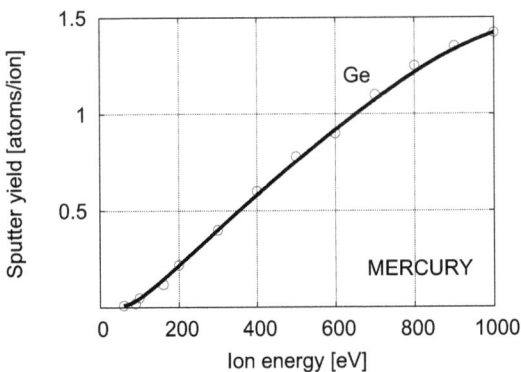

Figure 5.5: Sputter yield of Ge (111) caused by Hg$^+$ ion bombardement. The figure has be replotted due to quality issues of the original, source [54].

Chapter 5. Simulation of the sputtering process

Farren and Scaife [70] reexamined in 1968 the temperature dependence of the GaAs sputtering and found that the pattern corresponding to individual crystallographic orientations could be distinguished above 130 °C hereby reinforcing the statement made by Comas and Cooper. The work of Farren and Scaife was in so far important that it dealt with high Ar$^+$ impact energies 8 keV - 16 keV. It could be observed that for 16 keV Ar$^+$ impact onto the GaAs (111) surface, the low-temperature sputter yield of the amorphous surface was twice as high as the high-temperature sputtering of the crystalline surface (3.0 vs. 1.5 molecules/ion), see **Figure 3** in [70]. It is also interesting to observe the high-energy behavior of the sputter yield.

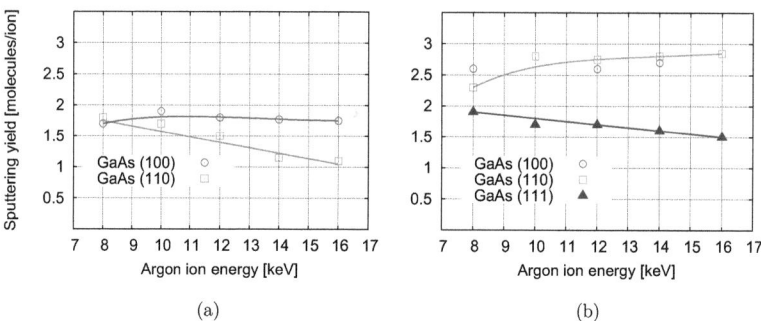

Figure 5.6: Sputter yield for GaAs at 250 °C, for (100) and (110) crystal orientation a), and at 20 °C - 70 °C for GaAs (100), (110) and (111) in b). Figures are taken from [70] and replotted due to quality issues of the original representation.

Figure 5.6 is replotted from reference [70] and presents the energy dependency of the sputter yield for three low-index crystal orientations of GaAs, namely the (100), (110) and (111) orientation in the energy range 8 keV - 16 keV. It can be observed that the sputter yield in this energy range is in most cases a decreasing function of the ion energy, corresponding to the high energy range of **Figure 5.3**. In one case, namely the (100) orientation, the sputter yield increases further between

Section 5.7. Validation and extension of the sputtering results

8 - 16 keV, this increase is, however, very small and the curve becomes almost horizontal at 16 keV, indicating that the maximum sputter yield for this crystal orientation has been reached.

Notice the similarity of the sputtering behavior of Ge in **Figure 5.5** with the lighter semiconductor representative silicon, obtained within the present work and presented in **Figure 7.5, Chapter 7**, also presented in [20]. The ratio of atomic masses for the Ge bombardment by Hg^+ ions is 0.7, while this ratio changes to 0.35 for the silicon bombarment by Ar^+ ions. Within the investigated energy range (0 - 1000 eV), the sputter yield of the Ge (111) surface is twice as high as it is for the Si (111) surface bombarded by Ar^+ ions. This is one of the first results for a semiconductor sputtering obtained in a low-energy interval 0 - 1000 eV.

6 Simulation of the nanoindentation

6.1 Introduction

Nanoindentation is a well suited method used for the evaluation of mechanical properties of materials, e.g., material hardness. Hereby, a hard indenter is pushed in a well controlled manner into the material under investigation while the force and the indentation depth are measured. On the micro scale this is enabled by using the nanoindenter as an experimental device as shown in **Figure 6.1**, for nanoscale indentations, higher resolution is required, therefore, an atomic force microscope (AFM) is used. For an overview of most commonly used indenter tips, the reader is refered to **Figure 6.2**.

All hardness measurements presented in this work are performed using UMIS 2000 (Ultra Micro Indentation System), produced by CSIRO (Commonwealth Scientific and Industrial Research Organisation), Sydney, Australia, see **Figure 6.1**. The UMIS 2000 has a force resolution of 1 μN and a distance resolution of 5 nm, hereby allowing measurments of surface near areas of bulk materials and thin films. During the indentation, the indentation depth and the force applied are measured. The analysis of the data is performed by the software Indent Analyser 1.6, developed by Asmatec GmbH (Advanced Surface Mechanics), Rossendorf, Germany. This allows an automatized determination of the hardness and Youngs modulus.

Since experimental nanoindentation has been applied for several decades for the evaluation of materials mechanical properties, it was

Chapter 6. Simulation of the nanoindentation

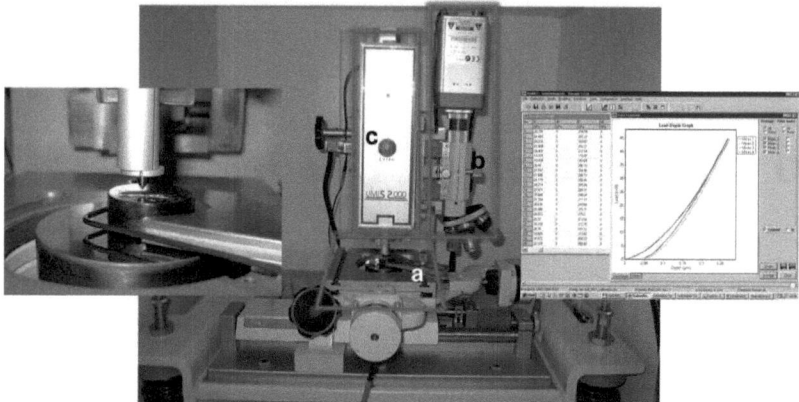

Figure 6.1: Schematics of the UMIS 2000 ultra microindentation experimental setup. In the central figure, the UMIS 2000 instrument is presented: (a) in x- and y-direction movable substrate holder, (b) the camera with 1500 magnification factor, (c) the measurement shaft holding the indenter tip. The smaller figure on the left shows in larger magnification the substrate placed under the indenter tip while the figure on the right shows the graphical user interface of the Indent Analyser 1.6 software. Pictures are taken by Ms. Janine Lichtenberg [71] at the IAM-AWP at KIT.

evident that interest into the numerical representation of processes during the nanoindentation will increase during time. It is especially the size of the indenter tip and the scales of the damage to the sample which makes this method very interesting for molecular dynamics simulation. In MD the nanoindentation is used for the simulation of the hardness analysis of bulk materials and coatings. System sizes may vary, depending on the demands onto the physical description and on computation time available but are most commonly in the range of 100 000 to a few millions of atoms.

Section 6.2. Nanoindentation of single crystals

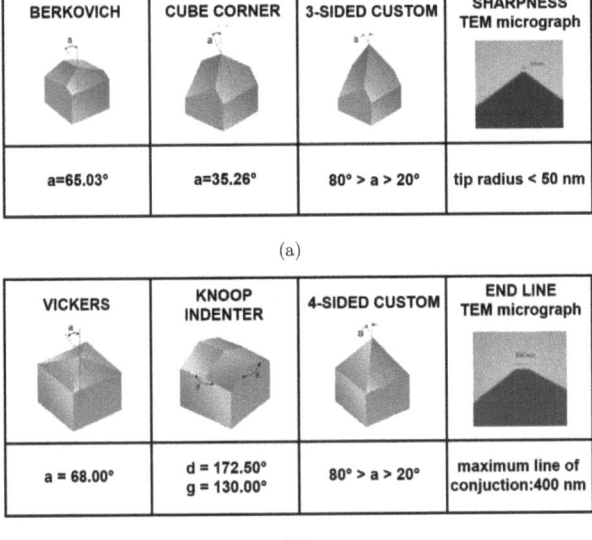

Figure 6.2: Most commonly used indenter tips, (a) three-sided (b) and four-sided. Berkovich and Cube Corner indenters are a traceable standard among the three-sided indenter tips, while Vickers indenter is the most prominent among the four-sided indenter tips, source [41].

6.2 Nanoindentation of single crystals

As the description of material properties of silicon by molecular dynamics (MD) was one of the driving forces of the development of this computational method, it was to expect that the simulation of the nanoindentation of silicon bulk material would be well described in the literature. Lin et al. [154] performed experimental and numerical investigation of the nanoindentation of the Si (100) surface by a diamond indenter tip. Covalent Si-Si bonds are represented by Tersoff potential [74, 76, 77], reproducing quite well the material properties of group IV elements. A Morse pair potential [122] was used to describe the interaction between the silicon substrate and the carbon atoms of the diamond indenter

Chapter 6. Simulation of the nanoindentation

tip. A complete rigidity of the diamond tip was assumed, a common simplification for the simulation of the nanoindentation, justified by diamonds the high material hardness (approx. 100 GPa). Lin et al. [154] adopted the method of constant displacement increment, evaluating the external load as the force acting on the indenter by the substrate atoms. Next to the load-displacement curve, which is the primary result of the nanoindentation experiment and simulation, they were especially interested in the stress distribution beneath the indenter tip and possible phase transformation of silicon due to these stresses. A pressure-induced phase-transformation of four-fold coordinated diamond silicon towards closer packed, higher coordinated forms of silicon such as metallic β-Si phase (Si-II at 11 GPa), rock-salt bcc-Si III and Si-XII was reported. Phase transformations were investigated by means of numerical analysis, such as coordination number investigation or evaluation of the radial distribution function (RDF) as well as by experimental methods such as cross-sectional transmission electron microscopy (XTEM) and Raman micro-spectroscopy [67, 81]. Similar results were reported by Cheong et al. [152].

Astala et al. [124] used the density functional based tight binding method to simulate nanoindentation of three low-index silicon surfaces by a sharp and a flat indenter tip. They showed that not only are the penetration depths smaller in the case of a flat tip, but also that the Si (111) crystal surface shows almost twice the hardness of Si (100) and Si (110) in the case of the sharp indenter tip, however values obtained by the flat tip show different results, see **Table III** in [124]. The resulting hardness values range from 28 GPa for the nanoindentation of a Si (111) surface by a flat indenter tip up to 89 GPa for the same crystal orientation but using a sharp indenter tip. These results are several times higher that those obtained by [154] and present simulation and experimental results and, thus, have to be taken with care. Systems sizes in the model of Astala et al. are very small, a cube of approx. 15 Å side length. Taking works of Lin et al. [154] and Astala et al. [124], questions which remain are the comparison of hardnesses of different Si

Section 6.2. Nanoindentation of single crystals

crystal orientations and the stress distribution beneath the indenter tip, both evaluated on a basis of a larger system.

Changing the material system from silicon towards silicon carbide further important works must be cited. Incorporating the Tersoff potential for multi-component systems [78], Noreyan et al. [16] performed a nanoindentation simulation of Si-terminated β-SiC (100) surface. The Tersoff potential was incorporated both for the description of Si-C-bonds within the bulk material as well as for the interaction of the diamond tip with the substrate SiC. Of special interest at this point was once again the transition from elastic to plastic behavior of the material, which is on the atomic scale represented by a phase-transformation of cubic β-SiC towards the 6-fold coordinated bcc rock-salt structure. Simulations were performed using the Verlet algorithm for the integration of motion [31] with the integration step of 0.175 fs and at different temperatures. The indenter was propagated by a constant displacement increment, the resulting indentation velocities were between 28,5 and 85,7 m/s. The critical indentation depth for the onset of elastic to plastic deformation was found to only weakly dependent on the indentation velocity. Further on, in the range of 100-450 K there was almost no dependence of the critical indentation depth on the temperature. Regarding the indenter size, it was found that the critical load needed for the phase-transformation decreases with increasing indenter width, which can be explained by the fact that stresses are more widely distributed beneath the indenter. Radial as well as angular distribution functions were investigated, the analysis of both suggest the phase-transformation towards rock-salt bcc structure of SiC. This is consistent with experimental observations, however, the exact pressure of the phase transformation is uncertain, due to limitations of the Tersoff potential which is not able to describe the ionic nature of the rock-salt structure.

In [66] Szulfarska et al. present the nanoindentation-induced amorphisation of β-SiC. Simulated system contained around one million atoms, corresponding to system dimensions of 309 · 309 · 108 Å in ($\bar{1}10$), (001) and (110) crystallographic directions, with indenter propagation direc-

Chapter 6. Simulation of the nanoindentation

tion parallel to (110) crystalographic direction. Szulfarska et al. adopted the Stillinger-Weber potential [48], initially developed for silicon, modified to describe the Si-C bond. Modeling of the nanoindentation process included a rigid Vickers indenter and a constant displacement increment, corresponding to the velocity of 100 m/s which is of the same order as in [16]. For the analysis of the simulation data, the authors analysed the load-displacement curve, angular distribution function and ring statistics. The angle distribution function, initially having a sharp peak at 109°, corresponding to the tetrahedra form, showed a broadening effect after the indenter load was applied, indicating the reaction of the substrate structure to the external load. In addition, the load-displacement curve showed a non-monotonic character, which was interpreted as an onset of plastic deformation. The relation between the drops in the load-displacement curve corresponded to subsurface deformations, which were identified as dislocation bursts, initiated at the indenter edges. For the detection of dislocations, a method combining the coordination number analysis and the ring statistics was used: every atom of the zinc blende structure is a part of overall 12 three-fold rings, an irregularity from this number indicates a dislocation. Phase transformation from the crystalline to the amorphous state was identified as the coalescence of different dislocations.

The same authors reported in 2007 again on the topic of nanoindentation of β-SiC using a larger system [61]. An analog approach was adopted as in [66], in addition, a comparison between three low-index crystal orientations of β-SiC, namely the (001), (110) and (111) orientation was performed. In their publication, the authors applied a self-developed potential [118], presented earlier in the same year [118], consisting of two- and three-body terms, therefore, not being a bond-order potential (due to explicit split up of many-body parts) and including effects of steric repulsion, partial charge transfer between atoms, charge-dipole and van der Waals interactions. Similar observations were reported as in [66] but slightly different hardnesses were measured, the highest value being 27,5 GPa for the SiC (111) plane. The increased

Section 6.2. Nanoindentation of single crystals

hardness of the SiC (111) orientation corresponds well with earlier observations by Astala et al. [124] for silicon. None of the authors compared the hardness orientation dependency of the β-SiC (Si) by other results, neither theoretical nor experimental. A possible explanation for this effect was proposed in [61]: In case of the Si (111) (and SiC (111)) orientation, the loading direction is parallel to one of the covalent bonds of the Si/SiC tetrahedra, while in the case of the (100) or (110) orientation, the substrate reacts predominantly by bond-bending. As a result, a decreased hardness is measured for the (100) and (110) orientation, since bond-bending is energetically favoured over bond lenght shortening. Pressure-induced phase-transformation of β-SiC was not observed in any case. This is due to the fact that the maximal hydrostatic pressure during the simulation never exceeded 50 GPa, which is half of the value of 100 GPa required for the β-SiC phase transformation towards the densely packed rock-salt structure as reported by Shimojo et al. in [47].

In the case of Si_3N_4 material system, one can also refer some previous experiences reported in the literature. The group around Kalia and Vashishta, this time with Walsh as the first author reported in 2003 about large scale MD simulations of the nanoindentation of Si_3N_4 by a rigid indenter [119]. Again, a self-developed interatomic potential, presented a few years earlier in [15], was used to describe chemical bonds within the Si_3N_4 system. A purely repulsive, steric potential was used for the description of the substrate-indenter interaction. It was the intention of the authors to analyse the hardness difference between the crystalline and amorphous state of Si_3N_4. For the crystalline state, the indentation was performed parallel to the {0001}-axis of the α-Si_3N_4 single crystal. A constant displacement increment was used during the simulation, values of the radial distribution function, angle distribution and stress analysis at every step of the simulation enabled the authors to make important statements about the processes during the nanoindentation. The evaluation of the data showed that the pressures and plastic deformation extended to a significant distance from the indented tip, originating from cracks at the crystal surface. The amorphous phase of Si_3N_4 showed lower hardness than the crystalline one, with values of 31,5 GPa for a-Si_3N_4 and 50,3 GPa for α-Si_3N_4 (0001). These values correspond in a good way to the experimental value of 31 GPa for α-Si_3N_4 (0001) reported by Suematsu in [62] or 40 GPa and 48 GPa reported by Chakraborty and Mukerji in [34], respectively, which represent the hardnesses averaged over several crystal orientations. A comparison of the hardness of a-Si_3N_4 was not made at that point due to lack of experimental values.

At this point, the results obtained by different groups within the last decade are summarized. Molecular dynamics (MD) simulations of the nanoindentation process of silicon, silicon carbide and silicon nitride were performed. In some cases, the hardness was measured by a sharp Vickers indenter or by a flat-punch indenter with a rectangular base. In addition, both theoretical ab initio investigations, as well as experimental measurements were compared. Results of MD simulations quite good reflect the experimental situations, while ab initio predictions of material hardness are too high. Silicon was reported as the weakest material, a phase transformation from the four-fold coordinated diamond structure towards the six-fold coordinated rock-salt structure was observed at the hydrostatic pressure of 11 MPa. A direct comparison with the silicon carbide, which has the same lattice structure by a smaller lattice parameter reveals a dramatic increase of material properties. The same phase transformation was observed, but at a ten times higher hydrostatic pressure, this is due to the stronger Si-C bond. Kalia et al. [119] reported a smaller hardness of the amorphous phase of silicon nitride in comparison to α-Si_3N_4, however α-Si_3N_4 shows twice the hardness of β-SiC.

6.3 Nanoindentation of coating systems

Next to hardness measurments on bulk materials, either in their amorphous phase [119] or as single crystal in different crystal phases [34]

Section 6.3. Nanoindentation of coating systems

and orientations [61, 124], a special emphasis needs to be given to the investigation of layered systems, such as presented later in this work. The reason for this attention to layered systems was given firstly by J. S. Koehler in his publication [87] from 1970. In [87] Koehler set the foundation for the design of a layered system whose mechanical properties, especially hardness, would exceed the properties of its individual components. Koehler proposed five principles for a system of alternating layers of a material A and B:

- Lattice parameters on the operating temperature of A and B are nearly equal. Two crystals should be grown epitaxially onto each other to avoid large strains at the interface.
- Thermal expansion coefficients of A and B should also be nearly equal so that temperature changes don't destroy the lattice match at the interface.
- The elastic constants of A and B should be as different as possible, this would require a large amount of energy to drive a dislocation from material A to material B.
- The bonding between the two materials should be large to prevent delamination.
- Both layer A and B should be thin, e.g., 100 atomic layers to prevent the generation of dislocations within individual A and B layers.

Considerations of Koehler may not at first seem relevant for the description of the nanoindentation as a method, however, they are the origin of all considerations later which lead to specific material design, together with the nanoindentation as a method to quantify mechanical properties of such systems. Nanolaminates consist, as proposed by Koehler, of nm thick, alternating layers of material A and B, with particle numbers in the range of several 100 000 atoms making them accessible for molecular dynamics simulations. Within the work group at IMWF, two publications dealt with the nanoindentation of multilayers,

Chapter 6. Simulation of the nanoindentation

both on the TiC/NbC(VC) material system. In [148] Sekkal et al. investigated in 2005 the supperlattice hardening effect at the TiC(110)-/NbC(110) interface by a nanoindentation simulation, see **Figure 6.3**. For this purpose they compared the hardness of a pure TiC (110) surface with the layered TiC(110)/NbC(110) system. Both systems were modelled as a cube of a side length of 15 nm and consisting of approx. 200 000 particles. A pyramid with an opening angle of $\alpha=142°$ was used to model the indenter tip. The hardness of a TiC (110) surface was determined to be 15.88 GPa which is close to the experimental finding of 18.68 GPa [11]. The hardness of the TiC(110)/NbC(110) layered system was evaluated to be 33.59 GPa, clearly indicating the increased hardness of this super lattice and corresponding to experimental observations made by Belger [13] on an example of a TiC/VC super lattice.

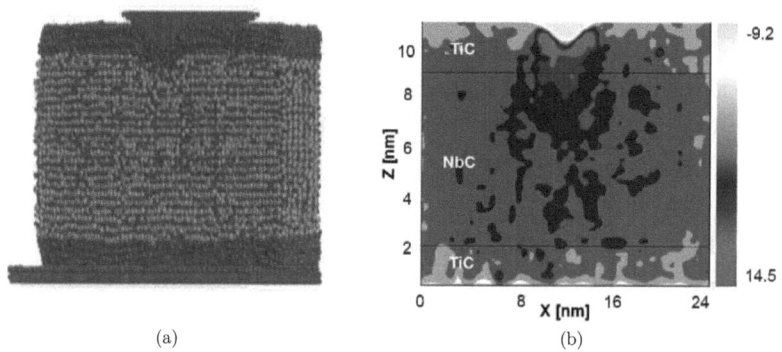

(a) (b)

Figure 6.3: Indentation state for the TiC(110)/NbC(110) interface at the indentation depth of approx. 10 Å and the resulting distribution of the σ_{zz} normal stresses, source [148].

In 2007 Kizler et al. investigated mechanical properties of TiC and NbC as well as of TiC/NbC layered system, [113]. Here, the autors extended the material system to several TiC/NbC bilayers and varied the carbon content in TiC and/or NbC (down to $Ti_1C_{0.9}$ and $Nb_1C_{0.9}$), see **Figure 6.4**.

Section 6.3. Nanoindentation of coating systems

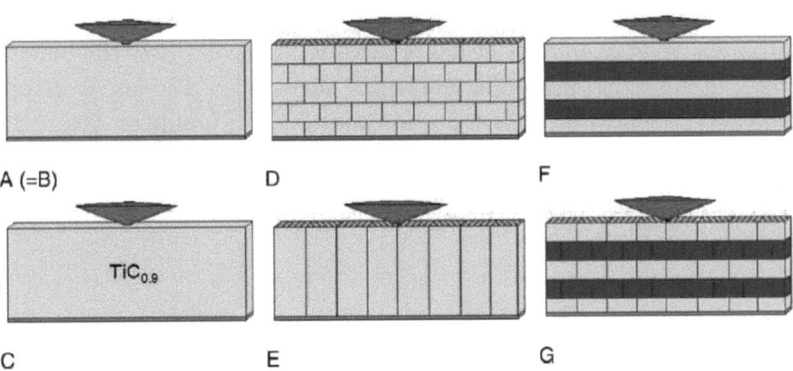

Figure 6.4: Principles of various simulation models, (A) TiC, (B) NbC single crystals, (C) Ti$_1$C$_{0.9}$ single crystal. TiC-crystal with grain and sandwich structures, tilt angle between the grain lattices approx. 35°, according to TEM observations from [13], symbolized at the top surface of model (D). TiC with columnar and tilted structure, tilt angles of approx. 35° (E), TiC/NbC sandwich structures made of single crystalline TiC and NbC columns with thicknesses of 4 nm each (E). (F) and (G) models represent TiC/NbC multilayers (TiC in yellow and NbC in blue) according to structures D and E, source [113].

Expertimentally, the investigation included the nanoindentation into the TiC/VC layered system, while in MD simulations, NbC was used instead of VC. This approximation was done because of the lacking of the EAM potential for VC at the time. A comparison of lattice constants of different carbides, a(TiC)=4.33 Å, a(VC)=4.16 Å and a(NbC)=4.57 Å, results in only 4 % difference for the TiC/VC pair and 5 % for TiC/NbC pair, hereby being compliant with the initial idea of Koehler [87]. Kizler et. al reported a nanohardness of 39 GPa for physical vapor (PDV) deposited VC and 42 GPa for PVD deposited TiC, in both cases a sapphire substrate was present. In the case of TiC/VC multilayers a general rule H=39.5 GPa+1.9 GPa/nm*λ was found, indicating a hardness increase propotional to the layer thickness λ. Similar observations were made by

Chapter 6. Simulation of the nanoindentation

MD nanoindentation simulations on the TiC/NbC layered system. In addition, a large amount of material amorphisation below the indenter tip was observed, remaining also upon unloading. The last statement will be verified by results on Si, SiC and Si_3N_4 in **Chapter 10**.

7 Results of MD sputtering simulations

In this chapter, the setup and results of MD simulations of the sputtering process of Si, SiC and Si_3N_4 by Ar^+ ions will be presented. In addition, a comparison with experimental results of Ar^+ etching, measured within the scope of this work, as well as with literature results will be made. A special emphasis will be given on the review of different theoretical and experimental predictions which could be achieved.

7.1 Computational details

The sputtering of different covalent materials by Ar^+ ions was simulated by classical molecular dynamics using both the commercial software package Materials Explorer 4.0 (ME, Fujitsu Ltd. [93]) and the program package IMD [73] to study the energy dependence of the sputter yield. For the modeling of covalent bonds within the target material, a Tersoff potential [45, 74, 75, 76, 77, 78] was applied, while the interaction with Ar^+ ions was described by the Ziegler-Biersack-Littmark (ZBL) pair potential [84]. The main advantage of ME is the user friendly graphical interface, allowing simulations without any programming knowledge, while its main disadvantage are constraints regarding computational time, because it runs on a single desktop machine. Once the relatively high amount of programming has been done, the IMD software package [73], running on a high performance computing cluster with thousands of CPU's, provides virtually unlimited computation possibilities, which allows the simulation of larger, physically more realistic systems. Nev-

Chapter 7. Results of MD sputtering simulations

ertheless, scientific results coming from both software packages possess a large amount of compliance, as it will be presented later on.

For the MD simulation of the sputtering process, periodic boundary conditions in the x-y plane and free boundary conditions in z-direction were adopted in order to minimize surface effects. The motion of the two bottom layers was restricted, hereby preventing a center of mass drift caused by the momentum transfer of the impacting Ar^+ ions. System dimensions were chosen to be $10 \cdot 10 \cdot 20$ unit cells, containing approx. 16 000 atoms. The chosen system size insured the complete transfer of the initial kinetic energy of Ar^+ ions to the target material and that the resulting collision cascades do not extend over the boundary of the simulated cell, preventing the cascade self-interaction due to periodic boundary conditions. At first, each sample was stabilized at 700 K using external temperature and pressure control by a Nose-Hoover thermostat and isotropic volume scaling. The integration time step was chosen to 1 fs. During the simulation, checkpoint files were written out every 100 femtoseconds, hereby providing thermally equivalent samples for the later statistical simulation of the sputtering process.

As described in **Chapter 5**, all processes within a collision cascade, caused by an impacting Ar^+ ion, are fast and occur far from thermodynamic equilibrium in a so called spike regime, characterised by high kinetic energies. In order to resolve these processes within an MD simulation, the integration step was set to $\tau = 0.1$ fs. For Ar^+ impacts, energies between 20 eV and 100 eV, in 10 eV steps were used, in order to determine the exact sputter threshold energy, between 100 eV and 1 keV, the energy increment was increased to 50 eV. The overall simulation time was chosen to be 40 000 time steps or 4 ps; within this time window all sputter relevant processes were assumed to be finished.

From this point on two strategies were pursued: firstly, distinct impact coordinates on the target surface were chosen, reflecting the crystal structure. This was done in order to distinguish different processes which may increase or decrease the back sputter yield, such as single knock-on effect, bond-breaking and channeling effect, described earlier in **Chapter 5, Sections 5.2** and **5.3**. The number of impact coordinates varied from 9 for Si (100) (see **Figure 7.1**) and β-SiC (100) (see **Figure 7.7(b)**) surface to 22 for the α-Si$_3$N$_4$ (0001) surface. Due to the fact that the experimental sputter yield is an average of numerous sputtering events, 50 equivalent simulations of the argon bombardment were performed, having the same Ar$^+$ impact energy and surface impact coordinate but slightly different atom coordinates within the target material due to thermal oscillations. Data such as Ar$^+$ trajectories, back sputter yield and number of forward sputtered atoms, defined as those moved into the negative z-direction for more than 1/4 of the lattice constant, equalling one atomic layer, were analysed and represented as the function of the Ar$^+$ impact energy. An analysis of back sputtered atoms regarding atom type and cluster size was performed. Using this procedure, it was possible to obtain a very accurate insight into the sputtering process.

The second strategy was oriented more onto the evaluation of the macroscopic sputter yield as measured in an experiment. Here, random target surface coordinates were chosen for Ar$^+$ impacts. In addition, several impacts onto the same crystal surface were performed, in between two impacts the sample was cooled down using the Nose-Hoover thermostat. The intention of this strategy was to represent high impact rates where only the excess energy in the form of the target temperature decays, but structural defects such as vacancies, amorphised regions caused by the ion impact remain in between impacts of two energetic ions. The purpose of this approach was to represent experimental results obtained by the group around Wehner and Anderson [52, 53, 54, 55, 56] since the target temperature was at the silicon annealing temperature of T_a=700 K (annealing temperatures for SiC and Si$_3$N$_4$ are not available and were assumed higher, in correlation to individual melting temperatures). This was done within MD simulation by the periodic alternation of the NVE and NVT ensemble, the first required to represent processes far from thermodynamic equilibrium as they occur within the collision cascade and the second one required to represent the system in ther-

Chapter 7. Results of MD sputtering simulations

mal equilibrium with its environment. By the described approach, it is possible to represent sputtering experiments on much longer time scale without a dramatic increase in computing time.

The described sputtering simulations were performed for 3 low-index crystallographic orientations, namely (100), (110) and (111) for Si and β-SiC and (0001), (10$\bar{1}$0) and (10$\bar{2}$0) for Si$_3$N$_4$. Again, argon trajectories, back sputter yield differentiated regarding atom type and cluster size were analyzed. In addition, a comparison of sputter yields of different crystal surfaces was made.

7.2 Sputtering process of silicon

Classical molecular dynamics (MD) has proved itself as a well suited method for the simulation of structural and mechanical properties of materials at the atomic scale, in the case of large systems (number of atoms larger than 1000). For the description of the sputtering process, the method of molecular dynamics is especially suited for lower impact energies, below 1 keV, at higher energies, the binary collision approximation as implemented, e.g., in TRIM [3], is more appropriate due to the spatial separation of the single collision cascades. In this section the sputtering process of Si as a model system will be dealt with while methods necessary for the description of the sputtering process of SiC and Si$_3$N$_4$ will be developed.

Figure 7.1 shows the top view onto the Si (100) surface as seen by an incident Ar$^+$ ion. Odd coordinates, nominated as K1-K7 represent individual surface atoms, starting from the top layer and descending a helically shaped way downwards into the silicon crystal structure. Hence, the coordinates K2, K4, and K6 represent chemical bonds of individual neighboring layer atoms. The coordinate K8 does not represent a chemical bond since it is a representation of the plane connection (here represented in white color) of the atom at K1 being in the top layer and the atom at K7 which is in the fourth layer counting from the crystal surface. The coordinate K9 represents a free, interatomic space in the

Section 7.2. Sputtering process of silicon

crystal lattice.

The four-corner K1-K3-K5-K7 square is periodically repeated on the Si (100) surface. In order to describe the sputtering process of the Si (100) surface, a description of the sputtering behavior within this four-corner is sufficient. For each impact coordinate (K1-K9) and each

Figure 7.1: Argon impact coordinates on a Si(100) crystal surface.

argon impact energy, ion penetration depth, back and forward sputter yield as well as the cluster sputter yield is analysed. Every data set is an average over 50 equivalent simulations, with the difference only in atomic coordinates of the target material due to thermal oscillations. In **Figure 7.2** the averaged back sputter yield is represented as a function of the surface impact coordinate for Ar^+ energies 700 eV - 1000 eV. As seen in **Figure 7.2**, the amount of back sputtered atoms is higher for if the incident ion hits a target atom directly. Also, the relative sputter yield decreases as the impact coordinate descends deeper into the crystal lattice. Argon impacts onto the coordinate K9 results in almost no sputter yield, the argon ion is channeled deep into the crystal structure making atomic collisions very unprobable. The resulting collision cascade is located deeper in the target material and overlaps with the crystal surface in very few occasions. Thus, the back sputter yield is practically zero for all impact energies. The overall scale

Chapter 7. Results of MD sputtering simulations

of the sputter yield ranges between 0.1 and 0.8 for all impact energies and all surface impact coordinates. The importance of the surface impact coordinate on the resulting sputter yield indicates that, for energies 20 eV - 1000 eV, sputtering is a surface dominated process. The governing mechanisms are therefore single knock-on and collision cascade, as described in **Chapter 5, Figure 5.2 (a) and b)**.

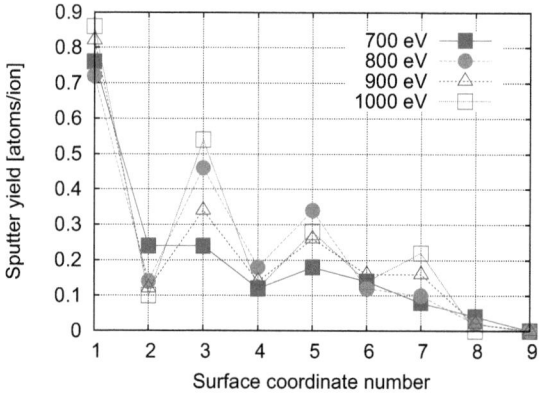

Figure 7.2: Si (100) sputter yield as a function of the argon energy and surface impact coordinate.

Next to the back sputter yield which is an information of direct practical use for the later deposition process, the quantification of the forward sputter yield provides an insight into the damage caused to the target crystal structure by the impacting ion. In **Figure 7.3** the forward sputter yield is presented as the function of the argon impact energy and the surface impact coordinate (K1-K9), solid lines show Bezier interpolations of the orignal data represented by points of the same color. For the definition of the forward sputter yield, crystal symmetry has to be taken into account. The diamond structure of silicon possesses four atomic layers in the {100} direction, hence, all atoms pushed by more that 1/4 of the Si lattice constant (one atomic layer) in the negative

Section 7.2. Sputtering process of silicon

z-direction are considered as forward sputtered atoms. All coordinates (K1-K8) representing Si lattice sites or interatomic space show similar forward sputter yield as the function of ion impact energy due to instant energy and momentum transfer, hereby resulting in higher forward sputter yield. The coordinate K9 shows the same tendency but approximately half of the value as the other impact coordinates, this is due to the absence of a direct argon-silicon impact.

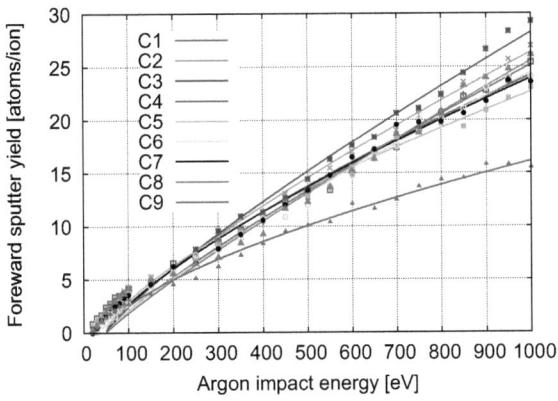

Figure 7.3: Forward sputter yield of silicon (100) as function of argon impact energy and impact coordinate as defined in **Figure 7.2**.

Finally, the analysis of the argon trajectory is presented. In **Figure 7.4** the penetration depth of an Ar^+ ion as a function of impact energy is presented. The averaging was performed over all impact coordinates and all 50 statistical cases. Coordinate K9 resulted in many cases in Ar^+ ion passing through the whole target length with only minor interaction with target atoms, this is formally known as the chanelling effect. Since in the case of the chanelling effect, the final coordinate of the argon ion at the end of the simulation depends not only onto the target thickness but also the overall simulation time, these cases were were not used and only cases in which a complete energy and momen-

Chapter 7. Results of MD sputtering simulations

tum tranfer from the Ar^+ ion towards the silicon target occured were taken into account. Two statements can be deduced from **Figure 7.4**, firstly, the penetration threshold energy of argon ions in silicon is approx. 70 eV, minor oscillations around 70 eV have their origin from the choice of the impact coordinate (direct hit on an atomic coordinate K1-K3-K5-K7 or easier ion penetration in the case of K9). Secondly, the penetration depth increases with the increasing ion impact energy. The increase is, however, not linear, the best fit was achieved with an exponent of 0.663. In [38], Humbird and Graves investigated ion-induced damage and annealing of silicon. As in this work, argon was chosen as a noble gas representative, however, different interatomic potentials were chosen for the simulation, the Stillinger-Weber potential [48] (instead of the Tersoff potential [77]) was adopted for the covalent Si-Si bond while a Molière pair potential [144] (instead of the ZBL-potential [84]) was selected to describe the Si-Ar^+ interaction. By measuring the thickness of the amorphous layer on the silicon surface, the authors quantified the damage caused to the silicon target. In case of the Ar^+ impact energy of 200 eV, the amorphous layer was 15 Å thick, which is in excellent agreement with the argon penetration depth, presented in **Figure 7.4**. Although, the penetration depth is practically zero below 70 eV, impacting Ar^+ can nevertheless cause a certain damage to the silicon target at the low energy range via collision cascade originating from a direct hit onto a silicon surface atom. Below 20 eV, Humbird and Graves observed only a sporadic damage to the 1-2 Å thick surface layer, which again corresponds to present observations made for silicon forward sputter yield, presented in **Figure 7.3**.

After a detailed analysis of Si (100) sputter yield, the focus moves towards the analysis of the macroscopic sputter yield of silicon. Due to the fact that macroscopic materials are typically polycrystals, consisting of variously oriented and sized single crystals, the free surface can be differently oriented. In order to represent this situation in a MD simulation with a limited number of particles and system dimensions, different single crystal orientations have to be taken into account. This

Section 7.2. Sputtering process of silicon

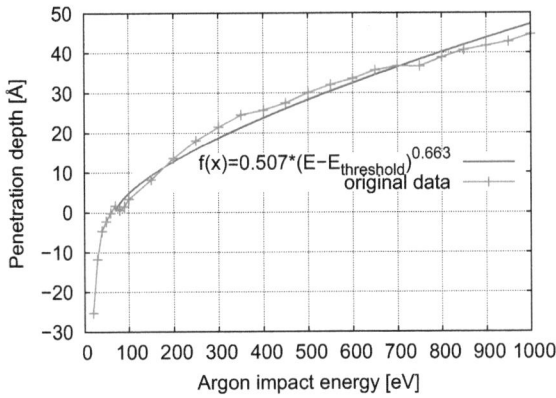

Figure 7.4: Averaged penetration depth of an argon ion as a function of impact energy.

step is unevitable for a reliable comparison and mutual validation of simulation and experimental results. In the case of silicon, 3 low-index crystal orientations, namely the Si (100), Si (110) and Si (111) were represented. **Figure 7.5** presents the calculated back sputter yield of silicon as a function of argon impact energy and crystal orientation. The three lower curves are calculated by IMD, while the pink curve represents the Materials Explorer calculation of the Si (100) sputtering. The first sputtering event occurs at an argon impact energy of 30 eV for the Si (111) crystal surface. Other sputter threshold energies are 40 eV for Si (110) and 60 eV for Si (100) crystal surface. In all cases, however, the highly statistical nature of the sputtering process has to be taken into account, so that the calculated sputter threshold energies present only the upper limit, individual sputtering events could theoretically be observed even for lower energies if instead of 50, more statistical cases were used. In addition, results of both software packages are compared by our experimental results obtained by ion etching of a 3 inch Si (100) wafer in an Ar microwave plasma (0.3 Pa, 1200 W, 400 °C) and mea-

Chapter 7. Results of MD sputtering simulations

surement of the etch depth by profilometry, presented in the light blue curve in **Figure 7.5**.

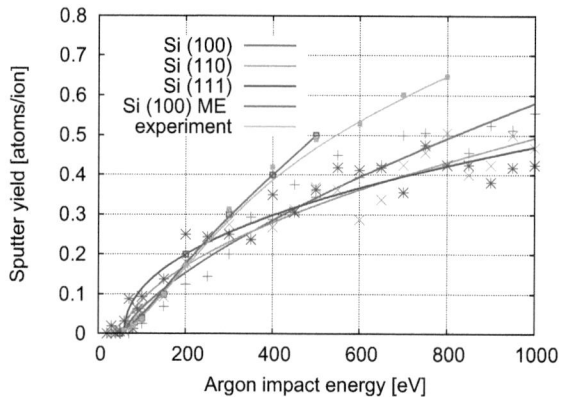

Figure 7.5: Sputter yield of silicon as function of argon impact energy calculated with IMD [73] for Si (100), Si (110) and Si (111) crystal orientation, by ME [93] for Si (100) and compared with experimental results of argon plasma etching.

A short overview of available data reveals the necessity of the presented investigation. In [140] Aoki et al. compared the sputter yield of Si determined by MD simulations for bombardment of Ar^+ ions in the energy range between 100 eV and 500 eV with MC simulations using the T-DYN code. They reported good agreement between both methods in this energy range; however, they made no comparison with any experimental data. In 1998, Kubota et al. [110] report MD simulations of Si bombardment by Ar^+ in the low energy range of 25 - 200 eV and comparison with available experimental data. Their study was in very good agreement with the experimental yields of Balooch et al. [94] and with results from MD studies of Barone and Graves [101]. Moreover, the calculated yield from [110] corresponded well with the yield estimates of Zalm [121]. In [120] Zalm measured the sputter yield of silicon

Section 7.2. Sputtering process of silicon

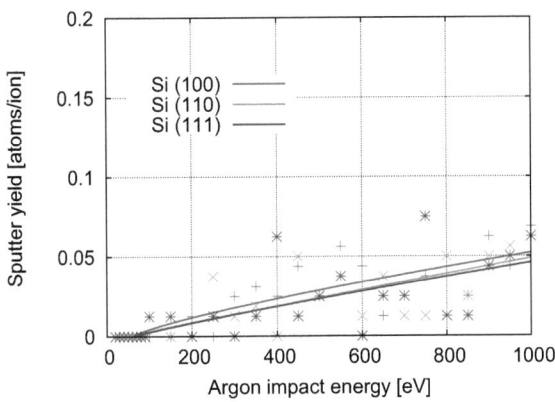

Figure 7.6: Silicon cluster sputter yield as function of argon impact energy calculated with IMD [73] for Si (100), Si (110) and Si (111) crystal orientation.

under normal incident angle Ar$^+$ bombardements for selected energies of 200 eV, 500 eV, 1 keV, 2.5 keV, 10 keV and 20 keV. The evaluation of the existing data shows that there is still a gap in the energy range between 20 eV and 1 keV where either experimental or simulation data are missing. Therefore, in present work, a comparison of MD simulations and experimental investigations of the sputtering process of silicon in the energy range of 20 eV-1000 eV, interesting for subsequent deposition process, is presented.

Next to the analysis of the overall sputter yield, an important information is the clustering tendency of back sputtered particles. **Figure 7.6** presents such an analysis as a function of argon impact energy. As seen in **Figure 7.6** the tendency of silicon to back sputter in form of clusters rather than single atoms is the same for all crystal orientations and compared with the **Figure 7.5** reveals that around 10 % of Si atoms are sputtered in the form of diatomic molecules. The amount of clusters containing 3 and more atoms is negligible. In [100] Timonova et al. performed MD simulations of the sputtering process of Si (100) surface by

Chapter 7. Results of MD sputtering simulations

argon ions at 45 ° impact angle and 500 eV impact energy. For the simulation, different interatomic potentials were used, special emphasis was given to the MEAM potential for silicon, presented by Baskes in [23]. The overall sputter yield of silicon ranged from 1.34 to 4.48, while the amount of back sputtered silicon dimers ranged from 4-16 %. Taking the impact angle into account, our own results correlate good both with the overall sputter yield as well as with the amount of back sputtered silicon clusters. It has to be stated however, that experimental mass spectrometry does not reveal any clustering tendency of the back sputtered Si-atoms. This is, therefore, either an artefact of the numerical modeling, or Si-clusters tend to desintegrate in further interaction with the argon plasma in the vacuum chamber and do not apper in a massspectrometer. Due to simplifications of the numerical model, presenting only one Ar^+ ion within the simulation cell, a correct answer to this question can not be given at this point.

7.3 Sputtering process of silicon carbide

In analogy to the sputtering process of silicon, sputtering of β-SiC was simulated by the method of molecular dynamics and validated by experimental results. At first, different impact coordinates on carbon terminated β-SiC (100) surface were defined in analogy to impact coordinates on the Si (100) surface and with respect to the zincblende crystal structure of β-SiC, see **Figure 7.7**. In case of carbon terminated β-SiC (100) surface, the uppermost layer consists of light, two-fold bonded carbon atoms, next layer consisting of silicon atoms etc. The analysis of the back sputter yield as the function of argon impact energy and impact coordinate is presented in the **Figure 7.8**.

In **Figure 7.8** slightly different presentation form is chosen than in the case of silicon sputter yield in **Figure 7.2**, the overal sputter yield of β-SiC (100) is presented in the energy range 20-1000 eV, this was done to show the continuous discrimination of the coordinate number 6. Coordinate 6 represents the arithmetic average between the location of

Section 7.3. Sputtering process of silicon carbide

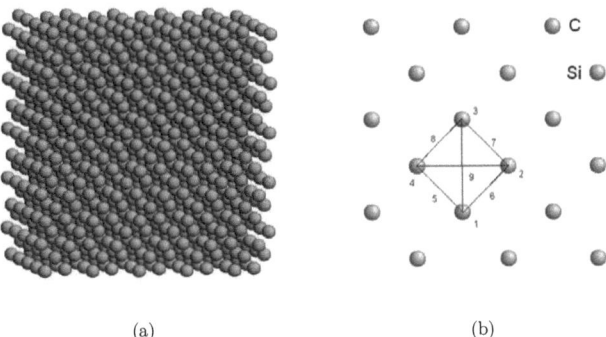

(a) (b)

Figure 7.7: Carbon terminated β-SiC (100) single crystal (a), visible are alternating layers of silicon (orange) and carbon (gray). Top view onto carbon terminated β-SiC surface with defined ion impact coordinates (b).

the top layer carbon atom and silicon atom in the next lower layer and is hereby the upermost covalent bond. Argon impact onto this covalent bond results directly in back sputtering, its value of 1.6 atoms per impacting Ar$^+$ ion is several times larger than the sputter yield averaged over all impact coordinates which is around 0.4 atoms/impacting ion at 1000 eV. Again, in the case where the initial argon trajectory is parallel with the channel within the interatomic free space (labeled as coordinate number 9) the resulting back sputter yield is practically zero. For other argon impact coordinates, the resulting sputter yield lies below 0.8 atoms/ion.

Penetration depth of argon ions in β-SiC shows similar energy dependency as for silicon bombardment, in average the first implantation of an Ar$^+$ into the SiC crystal lattice occurs at 90 eV, this threshold energy is slightly larger that in the case of silicon ($E_{threshold}(Si) = 70eV$) and can be directly correlated to the higher binding energy of β-SiC ($E_{bind}(\beta\text{-SiC})$=6.18 eV) in comparison to silicon ($E_{bind}(Si)$=4.62 eV) and the smaller lattice constant (a(β-SiC)=4.32 Å vs. a(Si)=5.5 Å). The argon penetration depth is in the case of β-SiC(100) bombardment around four times smaller than in the case of silicon and reaches around 12 Å for

Chapter 7. Results of MD sputtering simulations

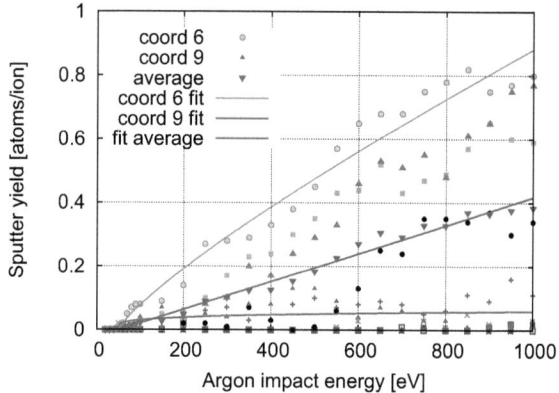

Figure 7.8: Back sputter yield of carbon terminated β-SiC(100) crystal surface as function of argon impact energy and surface impact coordinate.

1000 eV, again the average penetration depth for impacting free space is greater (here by a factor of 2) than for direct impacts.

Next to the back sputter yield and the penetration depth of argon ions in β-SiC, forward sputter yield of β-SiC is also important. As expected, this value depends on the argon energy and the impact coordinate onto the β-SiC (100) surface, see **Figure 7.10**. Here, it was distinguished between the direct impact of the argon ion onto one of the 4 atoms visible on the β-SiC (100) projection, the lateral impact and the argon ion penetrating through the channel in between the interatomic space (surface coordinate number 9). As depicted in **Figures 7.10** and **7.3**, the forward sputter yield of β-SiC (100) crystal orientation is around one half of the value of that for Si (100), this is directly related to the stronger covalent bond of Si-C compared to Si-Si. Again *free space impacts* cause less forward sputter yield as collisions are less probable and the argon ion looses kinetic energy more continuously. The average forward sputter yield of β-SiC (100) is directly proportional to the argon impact energy and has the value of 13 atoms/ion at 1000 eV impact

Section 7.3. Sputtering process of silicon carbide

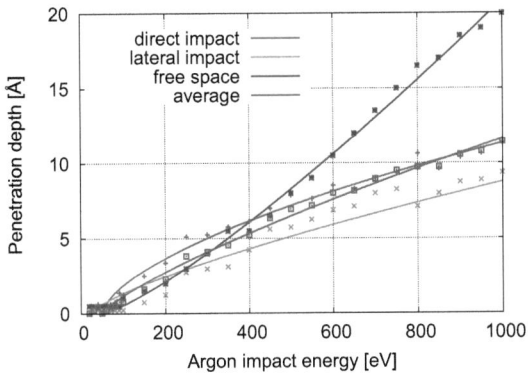

Figure 7.9: Argon penetration depth in β-SiC(100) as function of impact energy and impact coordinate.

energy (e.g. 25 atoms/ion in the case of silicon).

At this point the focus again moves towards the representation of the macroscopic sputter yield such as obtained in experiments by etching of a SiC target in argon plasma. The previously described method of SiC sputtering is insufficient for the description of the experimental situation. Up to now, only the back sputter yield of carbon terminated SiC (100) surface is calculated. In order to estimate the macroscopic sputter yield of SiC, back sputter yield of further crystal orientations has to be calculated. In addition, both carbon and silicon termination of the same crystal orientation has to be taken into account. It should be expected that, due to different atomic masses of silicon and carbon, different terminations show different sputter yields, e.g. carbon-terminated SiC (100) and silicon-terminated SiC (100). Also, the sputtering behavior of an amorphous SiC structure has to be taken into account. Farren and Scaife showed in [70] on the example of structure similar GaAs that an amorphous surface possesses a twice as large sputter yield as the crystalline structure. Due to diatomic unit cell of SiC, structural defects, caused by preceeding ion impacts will not heal in between individual

Chapter 7. Results of MD sputtering simulations

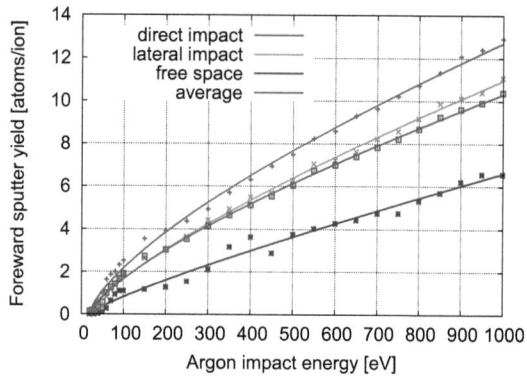

Figure 7.10: Forward sputter yield of β-SiC(100) as function of argon impact energy and impact coordinate.

ion impacts, causing an amorphisation of the target material during an experiment. This amorphisation could lead to a different back sputter yield as in the case of a SiC single crystal.

In order to overcome these limitations, the same approach as already adopted for the sputtering process of silicon was applied for β-SiC, with the goal to enhance the comparability of simulation results with experiments. Thus, sputtering of β-SiC (100), (110) and (111) surfaces was simulated by MD. In addition, between the silicon and carbon temination was distinguished in the case of β-SiC (100) and β-SiC (111). The results of the described simulations are presented in **Figures 7.12-7.21**.

The vertical scale (back sputter yield) in **Figures 7.12-7.20** is held constant at 0.7 atoms/ion for all β-SiC crystal orientations and terminations for better comparison. In all cases the overall sputter yield was analyzed in terms of atom type and cluster amount. In the case of a carbon terminated β-SiC (100) surface, **Figure 7.12** shows the highest sputter yield of 0.7 atoms/ion at 1000 eV, mostly out of carbon (silicon sputter yield at 0.1 atoms/ion at 1000 eV). Around 25 % of back sputtered particles release the target in form of clusters, mostly car-

Section 7.3. Sputtering process of silicon carbide

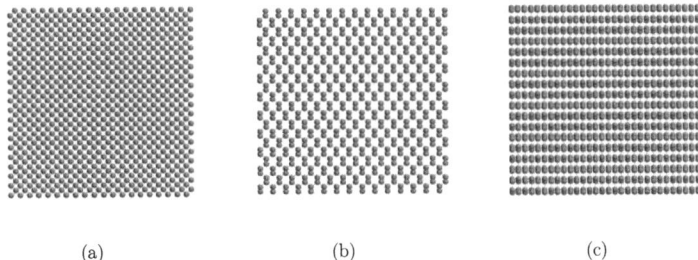

Figure 7.11: Top view onto 3 low-index β-SiC crystal surfaces, (a) SiC (100), (b) SiC (110) and (c) SiC (111). Visible are channels perpendicular to β-SiC (100) surface, causing diminished collision cascades withing the target materials and, herby, diminishing the back sputter yield.

bon dimers as it can be seen from **Figures 7.12** and **7.13**. The same orientation, but silicon termination delivers another sputter profile, the overall sputter yield is lower (0.55 atoms/ion at 1000 eV) with the same amount of carbon and silicon atoms. Silicon termination also decreases the cluster sputter yield by a factor of 4. This effect is predicted by the Sigmunds sputter theory, equation 5.1, described in **Chapter 5, Section 5.5**. Carbon, beeing a lighter compound of SiC shows higher sputter yield than the relatively heavy silicon atom.

Figure 7.16 and **7.17** show the sputter yield of β-SiC (110) orientation as a function of argon impact energy. In direct comparison to all other crystal orientations and terminations it becomes obvious that the dependency of the sputter yield on the ion impact energy is less pronounced (the increase is lower), it even shows a saturation-like dependency after approx. 700 eV. This is, however, nothing but an artefact of the non-sufficient statistics (smaller amount of argon impacts). The smaller energy dependency of the sputter yield in the case of β-SiC (110) orientation can be easily explained, by taking the crystal symmetry into account: β-SiC (110) orientation of the zincblende structure builds wide channels of free interatomic space which go through the whole system, see **Figure 7.11**. Their effect is similar to coordinate number 9 in

Chapter 7. Results of MD sputtering simulations

the simulation of single ion impact on the β-SiC (100) crystal surface, passing argon ions unhindered deep into the crystal structure, hereby diminishing the collision cascade and finally the sputter probability. The same effect was observed by Farren and Scaife in [70] for GaAs sputtering, a summarized representation of their observations is shown in **Figure 5.6, Chapter 5**. An interested reader will observe that in the case of β-SiC (110) orientation, the sputter yield of the lighter compound carbon is almost twice as large as of silicon, although both compounds are equally present at the crystal surface.

Section 7.3. Sputtering process of silicon carbide

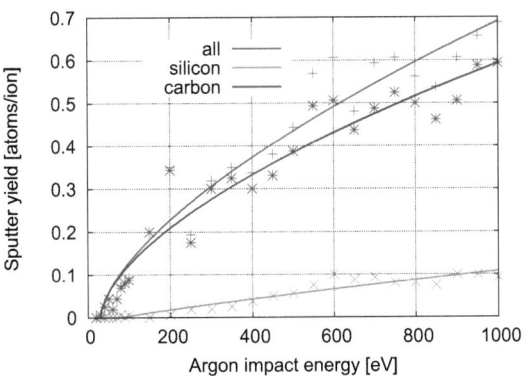

Figure 7.12: Differentiated sputter yield of carbon terminated β-SiC (100) as function of argon impact energy. The first sputtering event is observed at 30 eV argon impact energy, specifying the upper bound of the sputter threshold energy.

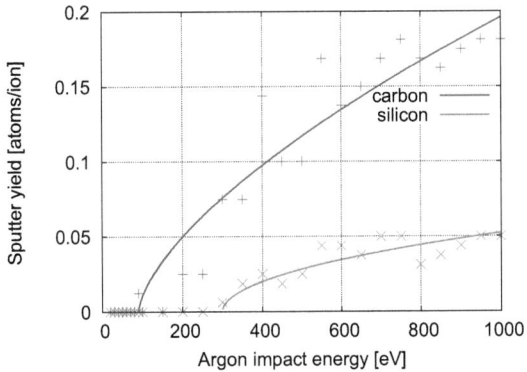

Figure 7.13: Differentiated cluster sputter yield of carbon terminated β-SiC (100) as function of argon impact energy.

Chapter 7. Results of MD sputtering simulations

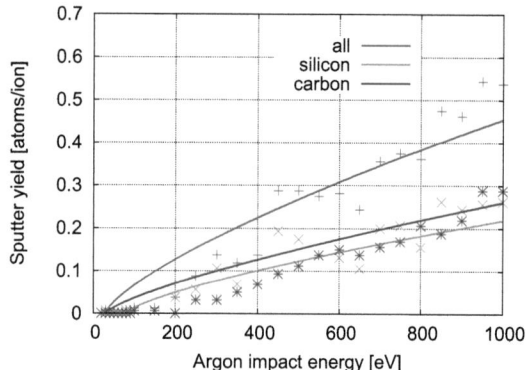

Figure 7.14: Differentiated sputter yield of silicon terminated β-SiC (100) as function of argon impact energy. The first sputtering event is observed at 30 eV argon impact energy, specifying the upper bound of the sputter threshold energy.

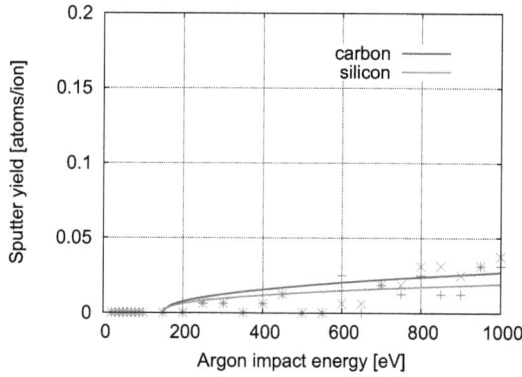

Figure 7.15: Differentiated cluster sputter yield of silicon terminated β-SiC (100) as function of argon impact energy.

Section 7.3. Sputtering process of silicon carbide

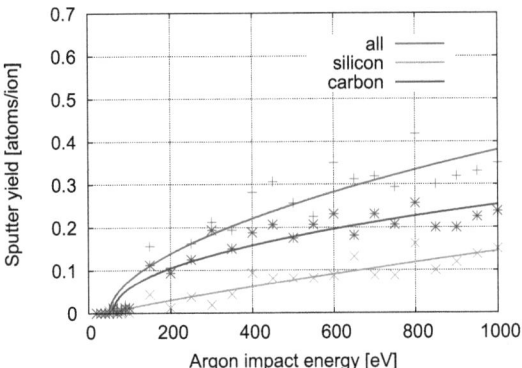

Figure 7.16: Differentiated sputter yield of β-SiC (110) as function of argon impact energy. The first sputtering event is observed at 50 eV argon impact energy, specifying the upper bound of the sputter threshold energy.

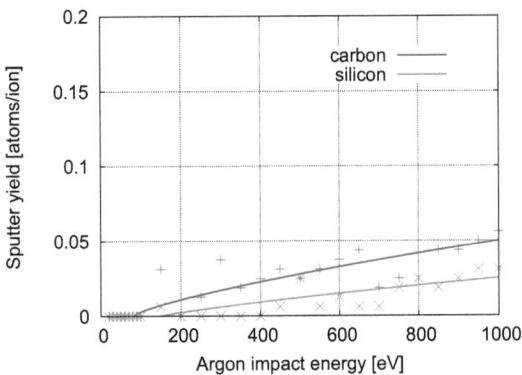

Figure 7.17: Differentiated cluster sputter yield of β-SiC (110) as function of argon impact energy.

Chapter 7. Results of MD sputtering simulations

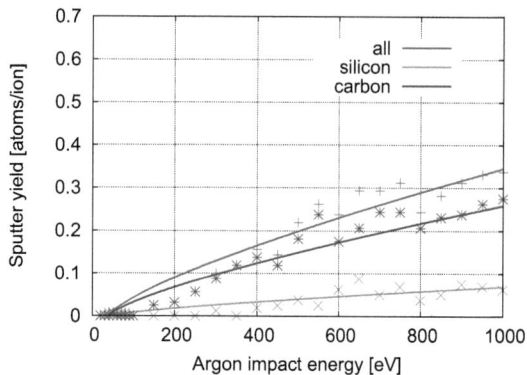

Figure 7.18: Differentiated sputter yield of carbon terminated β-SiC (111) as function of argon impact energy. The first sputtering event is observed at 40 eV argon impact energy, specifying the upper bound of the sputter threshold energy.

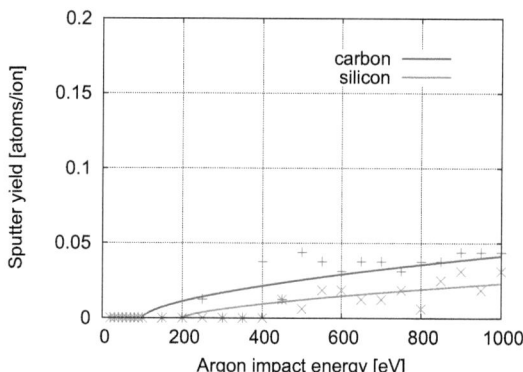

Figure 7.19: Differentiated cluster sputter yield of carbon terminated β-SiC (111) as function of argon impact energy.

Section 7.3. Sputtering process of silicon carbide

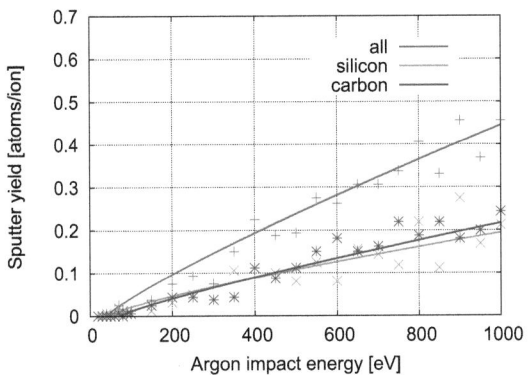

Figure 7.20: Differentiated sputter yield of silicon terminated β-SiC (111) as function of argon impact energy. The first sputtering event is observed at 40 eV argon impact energy, specifying the upper bound of the sputter threshold energy.

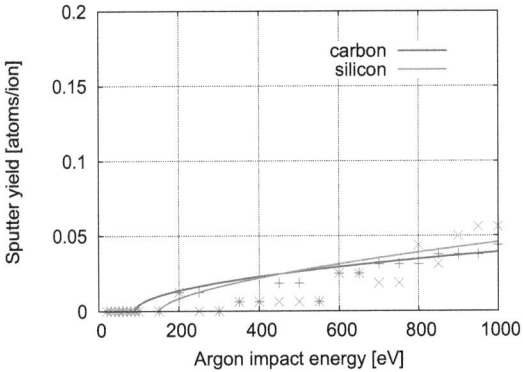

Figure 7.21: Differentiated cluster sputter yield of silicon terminated β-SiC (111) as function of argon impact energy.

Chapter 7. Results of MD sputtering simulations

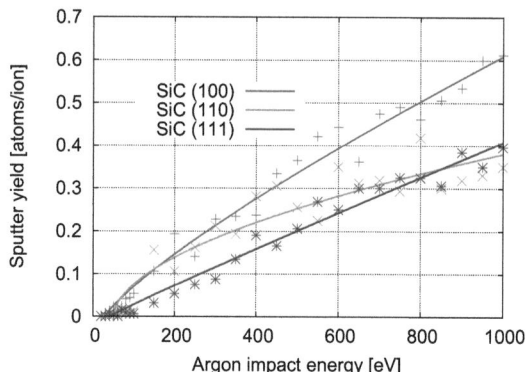

Figure 7.22: Comparison of calculated sputter yields of β-SiC for 3 low-index crystal orientations. In the case of β-SiC (100) and β-SiC (111) the average of the carbon and silicon terminations was used (see **Figures 7.12, 7.14** and **7.18, 7.20**).

7.4 Discussion and experimental validation

The averaged sputter yield from the IMD simulations in the energy range from 20 eV to 1 keV is compared in **Figure 7.23** with results from ME simulations, with TRIM [50] and TRIDYN [125] MC simulations and with values for the experimental sputter yield taken from the literature [19]. First sputtering events in MD simulations presented in this work are observed between 30 eV and 50 eV in IMD [73] simulations and 40 eV in MD simulations performed by Materials Explorer ME [93] simulation package. These results correspond well with the sputter thresold energy of 41 eV as reported by Kosiba in his PhD Thesis [125].

Experimental values for the low energy dependence of the sputtering yield for SiC by Ar$^+$ ions are rarely published. Mostly, there are experimental results for light projectiles like H$^+$, D$^+$, and He$^+$ [147] because of its importance in the nuclear fusion technology. Because in literature

Section 7.4. Discussion and experimental validation

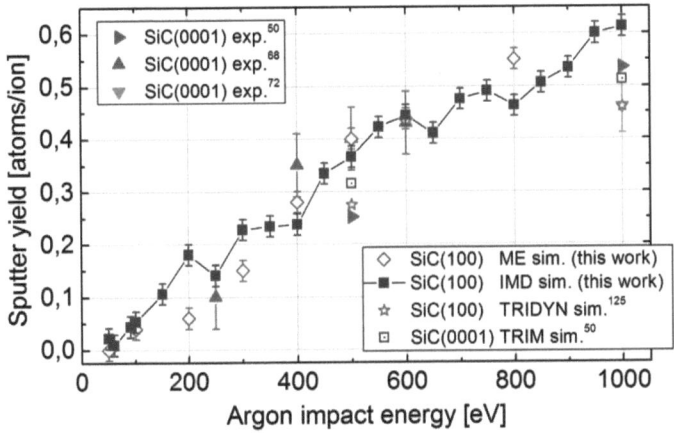

Figure 7.23: Comparison of experimental sputter yield values for SiC targets with TRIM [86, 155], TRIDYN and present MD simulations [19].

no experimental data have been found for the sputtering of a SiC (001) surface, the values for the sputter yield of SiC (0001) determined from target weight losses and measured by an in-situ microbalance [69], by the step height of mesa structures with a microinterferometer [72] and by Auger depth profiling [50] were chosen for comparison. It should be stated, however, that Petzold et al. [72] did not observe a dependence of the measured sputter yield on the polytype of the substrate, so the comparison with present simulations results is not straight forward.

7.5 Sputtering process of silicon nitride

As the last step of the sputtering process simulation, sputtering of α-Si$_3$N$_4$ by Ar$^+$ ions was performed. For this purpose, the method already tested for the sputtering process of silicon and β-SiC was adopted. The α-phase of silicon nitride which was used in MD simulations possesses a trigonal crystal structure, with lattice constants a=5.615 Å and c=7.766 Å as presented in **Figure 7.24**.

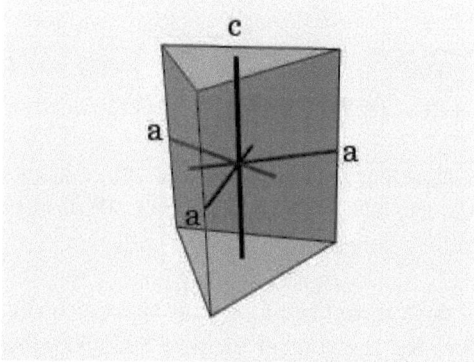

Figure 7.24: Schematics of the α-Si$_3$N$_4$ trigonal structure with lattice parameters depicted.

Subsequently, molecular dynamics simulations of multiple Ar$^+$ impacts on a α-Si$_3$N$_4$ target was performed and the resulting sputter yield was calculated. For this purpose, three low-index crystal surfaces of α-Si$_3$N$_4$ were bombarded by Ar$^+$ ions every 3 ps; in between individual ion impacts, the system was stabilized and the excess temperature brought by high energy Ar$^+$ ions was passed out of the system. Bombarded α-Si$_3$N$_4$ crystal surfaces were (0001), (10$\bar{1}$0) and (10$\bar{2}$0), as these surfaces are perpendicular to each other.

Section 7.5. Sputtering process of silicon nitride

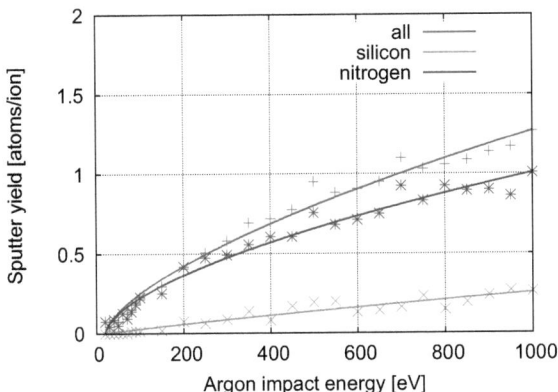

Figure 7.25: Differentiated sputter yield of α-Si$_3$N$_4$(0001) caused by Ar$^+$ impacts, in the low energy range up to 1 keV.

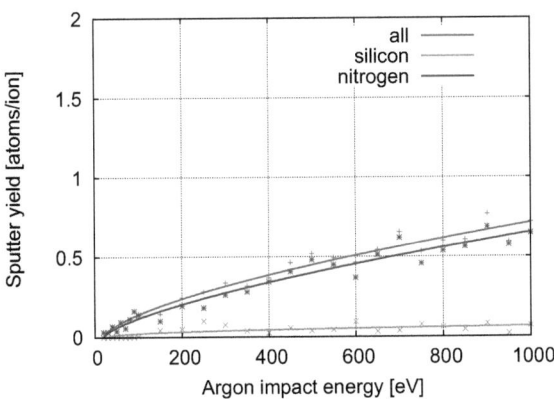

Figure 7.26: Differentiated cluster sputter yield of α-Si$_3$N$_4$(0001) caused by Ar$^+$ impacts, in the low energy range up to 1 keV.

Chapter 7. Results of MD sputtering simulations

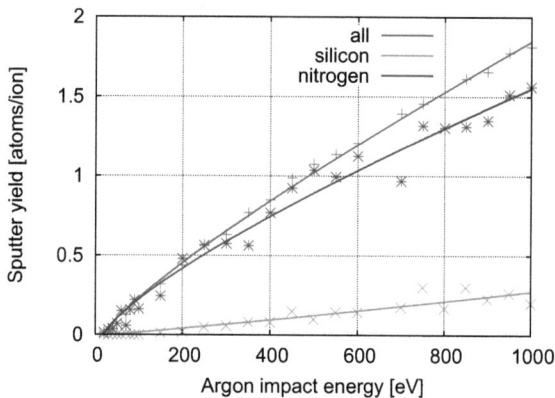

Figure 7.27: Differentiated sputter yield of α-Si$_3$N$_4$($10\bar{1}0$) caused by Ar$^+$ impacts, in the low energy range up to 1 keV.

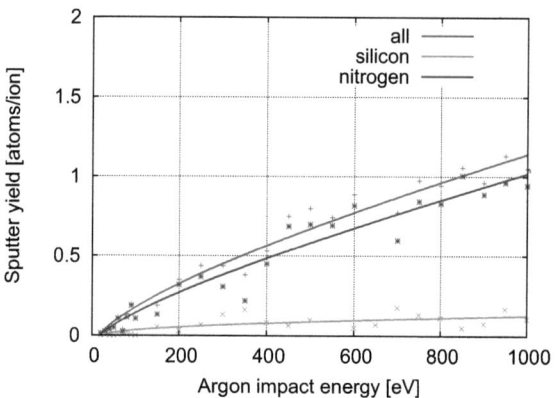

Figure 7.28: Differentiated cluster sputter yield of α-Si$_3$N$_4$($10\bar{1}0$) caused by Ar$^+$ impacts, in the low energy range up to 1 keV.

Section 7.5. Sputtering process of silicon nitride

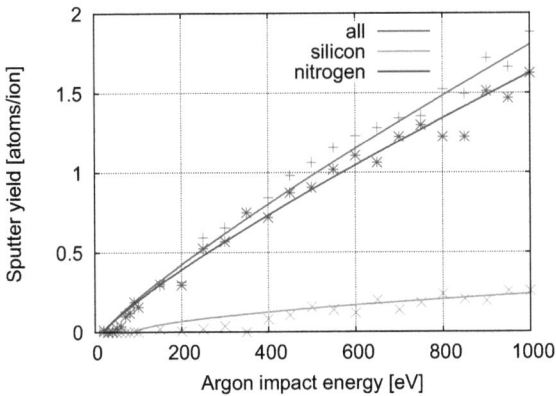

Figure 7.29: Differentiated sputter yield of α-Si$_3$N$_4$(12$\bar{1}$0) caused by Ar$^+$ impacts, in the low energy range up to 1 keV.

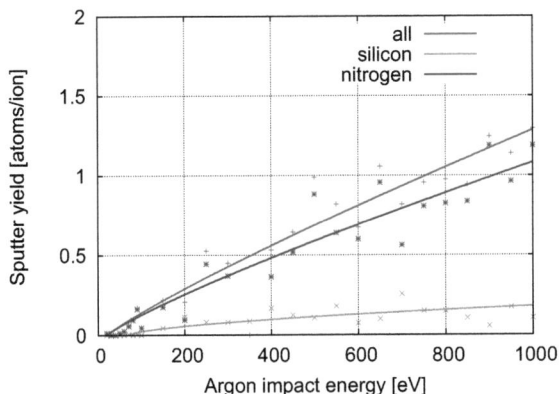

Figure 7.30: Differentiated cluster sputter yield of α-Si$_3$N$_4$(12$\bar{1}$0) caused by Ar$^+$ impacts, in the low energy range up to 1 keV.

Chapter 7. Results of MD sputtering simulations

7.6 Discussion and validation by literature results

The calculated sputter yields of individual crystal surface of α-Si_3N_4 are presented in **Figures 7.25-7.30**. The sputter yield of each crystal surface is presented in two diagrams, the first one containing the overall sputter yield and the sputter yield of individual atom species, while the amount of back sputtered clusters is presented in the second diagram. Additionally, back sputtered clusters were also analyzed by the means of particle type. In all cases, the cluster sputter yield was above 50 %, with approx. 90 % of back sputtered clusters being nitrogen molecules N_2. The fraction of back sputtered silicon atoms was very low, below 10 %. In addition, the silicon amount remains unchanged as the representation is changed from the overall sputter yield towards the cluster sputter yield, indicating that the majority of silicon atoms are back sputtered in the form of Si-N dimers.

In analogy to **Figures 7.5** and **7.11** for silicon and β-SiC, the orientation dependency of the sputter yield of α-Si_3N_4 is presented in **Figure 7.31**. It can be seen that α-Si_3N_4 ($10\bar{1}0$) and α-Si_3N_4 ($12\bar{1}0$) crystal surface have approx. 50 % higher sputter yield than α-Si_3N_4 (0001) surface. This discrepancy is directly related to the linear density of individual crystal orientations; α-Si_3N_4 (0001) orientation is parallel to the longer c-axis of the trigonal system, hence, the distance between the individual crystal layers is larger than in the case of two other orientations, making collisions less probable and, therefore, reducing the overall sputter yield.

Unfortunately, the literature lacks extensive results in the area of Si_3N_4 sputtering. To our knowledge, only two publications of Kim et al. in 2001 and 2006 have dealt with this topic. 2001 Kim et al. report in [36] about molecular dynamics simulations of the sputtering of amorphous Si_3N_4 surfaces by Ar^+ ions with impact energies of 100 eV and 200 eV and for different impact angles. The authors used explicit two- and three-body Stillinger-Weber based potential developed by Vashishta

Section 7.6. Discussion and validation by literature results

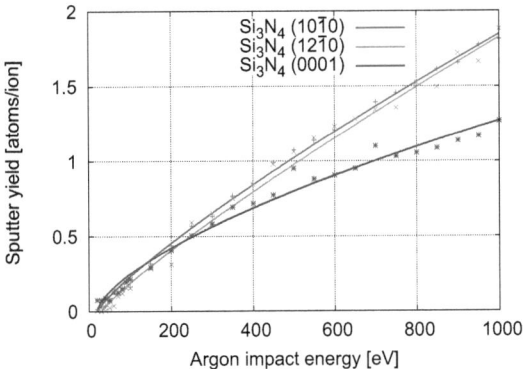

Figure 7.31: Comparison of calculated sputter yields of 3 low-index crystal orientations of α-Si$_3$N$_4$. Notice that individual crystal orientations are perpendicular to each other.

[15] to model the angle-dependent Si-N bond, while the interaction with the Ar$^+$ ion was described using the repulsive Moliere potential [144]. They found a maximum in the sputter yield between 60 ° and 75 ° and a nonstoichiometric surface atom removal (Si$_{\text{sput}}$:N$_{\text{sput}}$=1:2.5). These observations correspond well to the presented ones, although in our case the ratio of individual back sputtered atom types is larger, approx. 6 in favour of nitrogen. This can be attributed to the larger system in the present case (approx. 2000 particles in Kim et al. [36]) and 100 subsequent Ar$^+$ impacts onto the same surface. This would reduce the nonstoichiometry of back sputtered particles since the surface is continuously depleting on particles which are preferentially being back sputtered. In 2006 Kim et al. [37] extended the range of analysed material systems by modelling the sputter procces of a-Si$_3$N$_4$ and a-SiO$_2$ by He$^+$, Ne$^+$, Ar$^+$, Kr$^+$ and Xe$^+$ ions. Again, preferential sputtering of the lighter compound (N in the case of Si$_3$N$_4$ and O in the case of SiO$_2$) was observed. In addition, the overall back sputter yield of both compound materials was proportional to the mass of the impacting ion.

8 Basic concepts of experimental PVD

In this chapter, a short overview onto the method of physical vapor deposition (PVD) will be given. In addition, experimental investigation methods will be introduced and shortly described. The experimental part of this work was done by Priv.-Doz. Dr. Sven Ulrich, Dr. Carlos Ziebert and Dr. Jian Ye at the Karlsruhe Institute of Technology (KIT), Institute for Applied Materials - Applied Materials Physics (IAM-AWP). The purpose of this chapter to give a general overview onto the experimental part of this work rather than giving a detailed insight into individual investigation methods, for this the reader is referred to reference [25, 33, 99].

8.1 Experimental physical vapor deposition process

While the sputtering process is described in detail in **Chapter 5**, a more general view onto how this process fits in the concept of the physical vapor deposition is given here. Basic informations regarding plasma physics, direct current (DC) sputtering, high-frequency (HF) and magnetron sputtering (using both electric $\vec{E}$ and magnetic $\vec{B}$ field) will be provided.

8.1.1 Plasma physics

As plasma, normally a partially or fully-ionised gas is considered. Therefore, a plasma consists of neutral atoms n_0, ions n_i and electrons e and is globally neutral or *quasi-neutral*, meaning:

$$n_e = n_i \tag{8.1}$$

The amount of ionisation α of a plasma is given as the ratio of the number of ions (or electrons) to the overall number of atoms:

$$\alpha = \frac{n_i}{n_i + n_0} = \frac{n_e}{n_e + n_0} \tag{8.2}$$

For low-pressure plasmas between 10^{-3} and 10^{-2} mbar, the ionisation amount is between 10^{-4} and 10^{-3}. Within this work, Si, SiC and Si_3N_4 coatings are grown using Ar-plasma, which consists of single charged Ar^+ ions and electrons e^-.

8.1.2 Direct current DC sputtering

Igniting a plasma between two electrodes leads to a current of positively charged ions towards the target material, located at the cathode, while light electrons will be drawn towards the positively charged substrate, located at the anode. The kinetic energy of electrons in the potential field of several kV-s is sufficient to further ionize atoms on their way to the anode. Finally, before reaching the anode, the kinetic energy is low enough that no further ionization is possible, and, therefore, no luminescence can be observed. The space between the anode and the plasma is, therefore, called the Crook's dark space, a similar observation can be made at the cathode side. The Crook's dark space is normally around 0.1 mm thick.

On the cathode side, energetic Ar^+ ions hit on the target surface and cause a collision cascade in the target material. For all processes which occur along the ion trajectory, the reader is referred to **Figure 5.1** in **Chapter 5**. The amount of back sputtered material is highly parameter dependent, it can be estimated by Sigmunds sputter theory, **Section 5.5**. In the case of compound targets, it is often the lighter compound, which is preferentially back sputtered. This leads to an unstoichiometrical back sputter yield in the beginning, however, as the target material is depleted from the lighter compund, the back sputtering of the same becomes less probable. This will lead to stoichiometrical back sputter yield at a later stage.

8.1.3 High-frequency HF and magnetron sputtering

Next to DC sputtering, it is also possible to introduce a high-frequency electric field of approximately 13 MHz into the plasma. In this case, the high mobility of light electrons and the low mobility of relatively heavy ions, in our case Ar^+ ions, is used. The application of a high-frequency electric field leads to the development of a high potential decrease near the cathode within one oscillation period. Therefore, additional energy is applied on plasma ions, which impact onto the target material, neutralize there and with some probability lead to the sputtering of target atoms. The kinetic energy of ions is absorbed by the target material in the form of heat, making the cooling of the target necessary. Since back sputtered as neutral particles, the atoms of the target material propagate on a straight line, without any interaction with the electric field in the plasma and condensate on the substrate. On the substrate side, at the anode, the charge flux consists mostly of light electrons, which do not destroy the substrate crystal structure and only a small amount of positively charged, heavy Ar^+ ions, which can lead to back sputtering of the growing coating. Changing the potential on the substrate size may decrease or increase the amount of ions impacts, allowing impact rate control.

A special case of HF sputtering is called magnetron sputtering. Here, an additional magnetic field is introduced, acting on both electrons e^- and ions n^+. The Lorenz force $\vec{F}$ pushes the charged particles to move along circle trajectories with a radius r_{Larmor} and the frequency ω_{Larmor}

Chapter 8. Basic concepts of experimental PVD

perpendicular to magnetic lines $\vec{B}$:

$$\vec{F} = -e \cdot \vec{v} \times \vec{B} \tag{8.3}$$

$$r_{Larmor} = \frac{m_{ion}}{e} \frac{\vec{v}}{\vec{B}} \tag{8.4}$$

$$\omega_{Larmor} = \frac{\vec{v}}{r_{Larmor}} = \frac{e \cdot \vec{B}}{m_{ion}} \tag{8.5}$$

The magnetic field densifies the plasma, making collisions of particles more probable, ionizing the plasma further and hereby increasing the sputter yield. **Figure 8.1** gives a schematic overview onto processes occurring during magnetron sputtering.

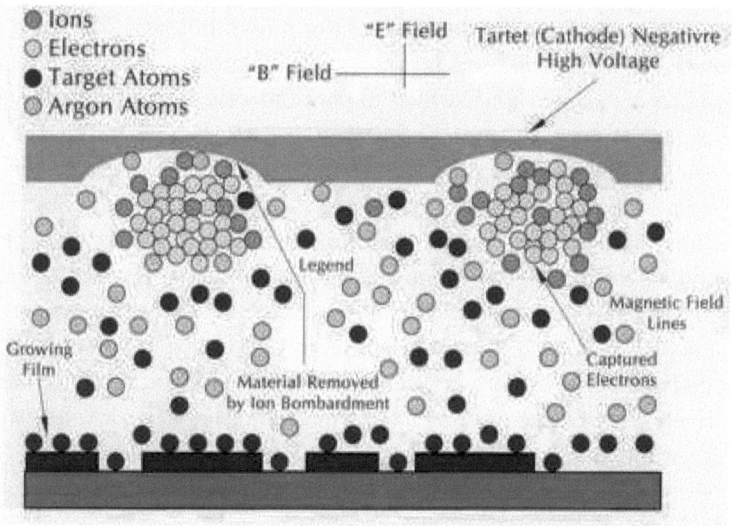

Figure 8.1: Schematic representation of the magnetron sputtering process, source [137].

8.2 Experimental characterisation methods

Subsequent to the deposition process, deposited coatings are investigated on their thickness, density and composition. Due to small scales, sophisticated experimental methods have to be applied. The following sections describe these methods and their basic principles. In addition, results obtained by these methods which are used in this work are named.

8.2.1 Measurement of the coating thickness

In dependence of the assumed coating thickness, several methods can be used to calculate the exact thickness of the coating.

On the lower range, it is possible to apply the surface profilometry. Prior to the deposition process a part of the substrate is covered by a kapton tape, preventing the coating growth at this region. Once the coating is deposited, the heigth of the step between the coating and the covered area of the substrate delivers the exact coating higth. It is often necessary to etch the substrate surface before the deposition process in order to increase the substrate roughness and hereby increase the substrate-coating adhesion. In this case, the etch depth has to be taken into account when meassuring the coating thickness, according to:

$$d_\text{coating} = d_\text{system} - d_\text{substrate} + d_\text{etch depth} \tag{8.6}$$

In the case of low coating thicknesess atomic force microscopy (AFM) can be applied in regions where the coating broke and the substrate beneath is visible.

For thicker coatings it is also possible to use the callote polishing method. By this method, a sphere of a known radius is polished into the substrate-coating system until the substrate becomes visible. An exact optical measurment of the formed ring, with known callote sphere radius reveals the exact coating thickness. In addition, the coating thickness can be measured by an optical or a scanning electron microscope (REM)

Chapter 8. Basic concepts of experimental PVD

on the breaking edge or by X-ray reflectometry (XRR).

According to [33], the results of all three stated methods are in accordance to each other, allowing the choice of a single method, fitting the best to the system under investigation.

8.2.2 Density analysis

Next to the coating thickness, it is also necessary to know the exact density of the deposited coating. This can be done by X-ray reflectometry (XRR) at a grazing angle. Hereby, the coating is irradiated by X-ray beams, which reflect both on the coating surface as well as on the substrate-coating interface. The resulting optical path difference of the two reflected beams lead to interference, its intensity is a function of the grazing angle. The intensity distribution obtained in this way can be simulated by a recursive equation of the reflection coefficient, according to Parrat [92], based on the Fresnel equation. The critical angle for the total reflection is a function of the coating density. This approach can also be extended onto XRR density analysis of multi-layer coatings, in this case, interfaces between individual coating layers are responsible for the optical path difference of the reflected beams. For coating thicknesses between 80 nm and 120 nm, XRR can be used to measure the coating thickness $d_{coating}$, coating density $\rho_{coating}$ and surface roughness R_α.

8.2.3 Chemical composition

For the investigation of the chemical decomposition of the coating material, the electron microprobe (EPMA [1]) as well as Auger electron-microscopy are used.

In the case of EPMA, high energy electrons are capable of knocking out near shell electrons of the material under investigation. The so emerging electron vacancies are then filled by electrons from the outer shells. The energy difference can be observed as X-ray radiation of a specific wave lenght. EPMA allows measurments of elements from boron to plutonium down to levels as low as 100 ppm.

Auger electron-spectroscopy is a surface sensitive analysis method. It is similar to EPMA, but has a distinct difference to it. Instead of detecting the radiation of the outer shell electron filling in the electron vacancies in the lower shells, this radiation is rather absorbed by a third electron, so called Auger electron. The Auger electron is, therefore, released from the material with a specific kinetic energy, depending on the previous electron-vacancy process and is a characteristic of a specific atom type. In principle, AES makes it possible to detect all chemical elements with at least 3 electrons, hence, hydrogen and helium can not be detected by AES.

8.2.4 Microstructure and bond-type analysis

X-ray diffractometry XRD is a destruction-free investigation method, by which structures in crystalline materials can be analysed. Often, an XRD in Bragg-Brentano alingment is used [99]. Irradiating the sample by an X-ray beam of wave lenght λ and under the angle θ leads do diffractions onto individual lattice planes of the crystal structure according to Bragg's equation:

$$2 \cdot d \cdot \sin \theta = n \cdot \lambda \tag{8.7}$$

where n is an interger number. Using XRD, it is possible to analyse periodic, crystal structures, while amorphous structures remain undetected (*X-ray amorphous*).

In the case of Electron Energy Loss Spectroscopy or EELS, the method is used to characterize bond-hybridisation (sp^2 or sp^3) in carbon. The sample is subjected to an electron beam in the energy range around 200 keV. From the energy loss within the sample, caused by specific binding states, one can deduce the bond-types within the material.

For the exact analysis of the microstructure, high resolution trans-

Chapter 8. Basic concepts of experimental PVD

mission electron microscopy TEM is used. Here, a sample is subjected to high energy electrons (approx. 300 keV). The distance between individual lattice planes d can be determined according to:

$$d = \frac{\lambda L}{R} \qquad (8.8)$$

where λ is the electron radiation wave length ($\lambda(300 \text{ keV})=1.968$ nm), L is the distance between the photo plate and the sample under investigation and R it the radius of the resulting interference rings on the photo plate. In principle, XRD and high resolution TEM are similar methods, as is their purpose. An advantage of the high resolution TEM is certainly the varaibility of the wave lenght, obtained by changing the electron impulse, via de Broglie relation:

$$\lambda_e = \frac{h}{p_e} \qquad (8.9)$$

8.2.5 Fourier Transformation Infrared Spectroscopy

Fourier Transformation Infrared Spectoscropy FTIR enables the analysis of present bond configurations. Hereby, a beam of infrared light is used to analyze the sample, the wave number ranges from 400 to 10 000 cm^{-1}. The wave numbers of the infrared light correspond to differences of individual energy spectra, both rotational and oscillatory, therefore, depending on the bond configuration in the system, specific wave numbers of the incoming infrared light are absorbed according to the Lambert-Beer law:

$$\frac{I_d}{I_0} = e^{-\alpha_\nu d} \qquad (8.10)$$

I_0 is the intensity of the incoming light, I_d is the intensity of the light trasnmitted through the specimen, α_ν is the co-called extinction coefficient corresponding to the intensity which is lost after passing through

Section 8.2. Experimental characterisation methods

1 nm coating and d is the coating thickness.

8.2.6 Raman spectroscopy

The basic principle of the Raman spectroscopy is the photon-phonon interaction. A difference to FTIR is the increased activity of homonuclear bonds, such as Si-Si and C-C. Because of this, Raman spectroscopy is often used a complementary intestigation method to FTIR.

8.2.7 Analysis of coatings mechanical properties

Coating materials, such as SiC and Si_3N_4 possess desireable mechanical properties in their bulk form, which makes them attractive for the deposition on silicon and on hard metals for, e.g., corrosion protection. Mechanical properties such as hardness and Young's modulus can be measured by nanoindentation. Since both the experimental and the simulation aspect of this method are extensively described in **Chapters 6** and **10**, the description of the evaluation of mechanical properties is restricted here to the analysis of residual stresses.

In general, for the deposition process, two different materials are used, both of them having different material properties, lattice parameters etc. in their bulk forms under certain conditions such as temperature and pressure. In addition, the PVD deposition process includes high energies and rates of film forming particles, making residual stresses also process dependent. Coatings deposited by magnetron sputtering show in general compressive residual stresses, while those grown by evaporation deposition show in general tensile stresses, the difference of stresses originates from different kinetic energies of the film forming particles [99]. Increasing the plasma pressure, e.g., from 1 Pa to 10 Pa will lead to tensile stresses within the coating, another possibility to change the stress state of the coating is the variation of the substrate bias voltage. Residual stresses are present only in the plane orthogonal to the film growth and lead to the bending of the substrate material. The Jaccodine

Chapter 8. Basic concepts of experimental PVD

relation enables one to evaluate the residual stresses from the bending h_B over the measured lenght L:

$$\sigma = \frac{E}{1-\nu} \cdot \frac{d_{sub}^2}{6 \cdot d \cdot R} = \frac{E}{1-\nu} \cdot \frac{4}{3} \cdot \frac{d^2 \cdot h_B}{L^2 \cdot d} \qquad (8.11)$$

with

$\sigma =$ Coating residual stress [GPa]

$E =$ Young's modulus [GPa]

$\nu =$ Poisson's ratio

$d =$ substrate thickness

$d =$ coating thickness

$L =$ measured lenght

$R =$ curvature radius

The Jaccodine equation is correct only for low bendings, when the curvature radius R is

$$R = \frac{L}{4}\left(\frac{L}{2h_B} + \frac{2h_B}{L}\right) \approx \frac{L^2}{8h_b} \qquad (8.12)$$

for $L \gg h_B$.

9 MD simulation of the deposition process

Subsequent to the simulation of the sputtering process, the deposition process of a silicon substrate by silicon, SiC and Si_3N_4 was simulated. In order to achieve this, a cube-shaped silicon sample was used as substrate and was kept constant at different temperatures, ranging from room temperature to 1500 K. This was done in order to investigate the temperature dependence of the coating structure. On the substrate side, the crystal orientation was another input parameter, silicon single crystals in (100), (110) and (111) crystal orientation were used in order to investigate how the crystal orientation of the substrate material affects the coating structure. On the side of the film-forming particles, it was the energy of the film-forming particles and the particle-dependent deposition rate which directly influenced the coating structure.

9.1 Description of the methodology

For the simulation of the deposition process, the method of molecular beam epitaxy (MBE) was used and will be described in more detail in the following. The MBE allows the simulation of the atom deposition onto a substrate. Using temperature control as described, was is possible to model the deposition process at different substrate temperatures, the same can be achieved in experiment by cooling or heating of the substrate in the vacuum chamber. In order to achieve a constant deposition rate, particles were created at fixed height above the substrate and at random X- and Y-positions. For each particle type, both the deposition

Chapter 9. MD simulation of the deposition process

start time as well as the deposition rate can be varied independently. This approach makes it possible to model a subsequent deposition of different material systems leading to multi-layer coatings as well to change the deposition rate of an individual particle type during the simulation resulting in the deposition of gradient coatings. The later ones will be described in this chapter. At a constant deposition energy, the created particles are given a momentum towards the substrate where they interact with the substrate surface. The incident particles are either adsorbed by the substrate surface or reflected, depending on the incident energy, particles can also penetrate into the substrate bulk, causing a collision cascade and possibly a back sputtering of the substrate or of already deposited coating material. All of the stated scenarios are possible during a real experiment. A simple schematics of MBE, as presented in [73], is given in **Figure 9.1**.

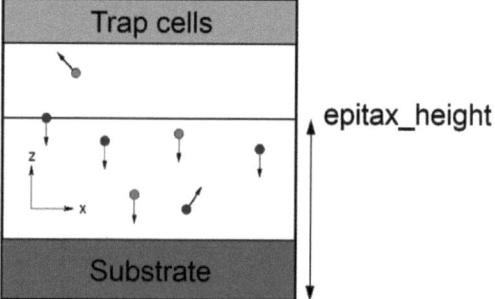

Figure 9.1: Schematic representation of molecular beam epitaxy (MBE) simulation method, source [73].

9.2 Deposition process of silicon on a Si substrate

As a first step towards a successful deposition of SiC and/or Si_3N_4-coatings, MD simulations of silicon deposition onto a silicon substrate is performed in order to investigate basic influences of deposition parameters onto the coating structure.

For the investigation of the temperature dependence on the coating formation, the silicon substrate was kept constant at two different temperatures, namely at 500 K and 1000 K. This idea is based on the work of Gawlinski et al. [44], who again were motivated by previous works of Gossmann and Feldmann [40, 57, 58] in the mid 80-s of the last century. Gossmann and Feldmann investigated the temperature dependency of silicon and germanium coatings on silicon substrates in different crystal orientations and postulated both an epitaxial transition temperature $T_{epitaxial}$, below which only amorphous coatings can be deposited as well the limiting coating thickness $h_{epitaxial}$ for the growth of epitaxial silicon coatings in dependence on the deposition temperature. The substrate orientation was also varied, Si (001), (110) and (111) crystal orientations were used in order to investigate the dependence of the coating structure on the substrate orientation.

Deposition rates of silicon atoms were 0.1 particle/ps, or 1 particle/ps, while their kinetic energies were set to 1.5 eV, 3 eV, 6 eV and 10 eV respectively. The momentum of incoming particles was set perpendicular to the substrate surface, this resembles the experimental situation by far.

9.2.1 Silicon coating structure vs. deposition parameters

Figure 9.2 shows homoepitaxial deposition of silicon on Si (100) as a function of substrate temperature. The presented Si-substrate/Si-coating systems were obtained by a previousely described method of

Chapter 9. MD simulation of the deposition process

MBE and analyzed in terms of the potential energy and stress distribution. In **Figures 9.2(a)-9.2(c)** an amorphous silicon coating is depicted, deposited at 500 K. The amorphous structure of the coating is visible in **Figure 9.2(a)**. In **Figure 9.2(b)**, potential energies of individual atoms within the Si-substrate/a-Si-coating system, as calculated by the MD simulation are analysed and presented in terms of different colors. Low potential energy, corresponding to four-fold coordinated silicon bulk atoms in an equilibrium position is presented in red color, while silicon atoms within the amorphous network of the Si-coating a have higher potential energy and are presented in orange, yellow, green and blue color. During the simulation of the deposition process, the stress tensor of each atom is calculated according to the Virial theorem [91]. For the presentation, von Mises equivalent stress is chosen, it can be calculated from the stress tensor:

$$\sigma_{vM} = \sqrt{\frac{1}{2}[(\sigma_{xx}^2 - \sigma_{yy}^2) + (\sigma_{yy}^2 - \sigma_{zz}^2) + (\sigma_{zz}^2 - \sigma_{xx}^2) + 6(\tau_{xy}^2 + \tau_{yz}^2 + \tau_{zx})]} \quad (9.1)$$

with individual σ_{ii} being normal stresses, τ_{ij} shear stresses and σ_{vM} the resulting von Mises equivalent stress.

In **Figure 9.2(c)** the distribution of von Mises stresses σ_{vM} within the Si-substrate/a-Si-coating (amorphous silicon coating) is shown, visible is the low stress state of the silicon substrate (red color) and high von Mises stresses within the amorphous network of the coating. Below, in **Figure 9.2(f)**, a typical Si-substrate/c-Si-coating (crystalline silicon coating) as a result of high temperature deposition at 1000 K is presented. The coating shows ordered structure, a direct comparison of the potential energy distribution in **Figures 9.2(b)** and **9.2(e)** indicates that most of silicon atoms in the crystalline coating structure occupy equilibrium positions. This is possible due to a higher average kinetic energy within the system, enabling the surmounting of potential barriers which occur during the deposition process. Finally, **Figure 9.2(f)** presents the distribution of von Mises stresses within the crystalline Si-

Section 9.2. Deposition process of silicon on a Si substrate

substrate/Si-coating system. On the contrary to the amorphous coating in **Figure 9.2(c)**, both the substrate-coating interface as well as the overall silicon coating show smaller von Mises stresses, indicating high substrate-coating adhesion and a lower stress state of the coating itself.

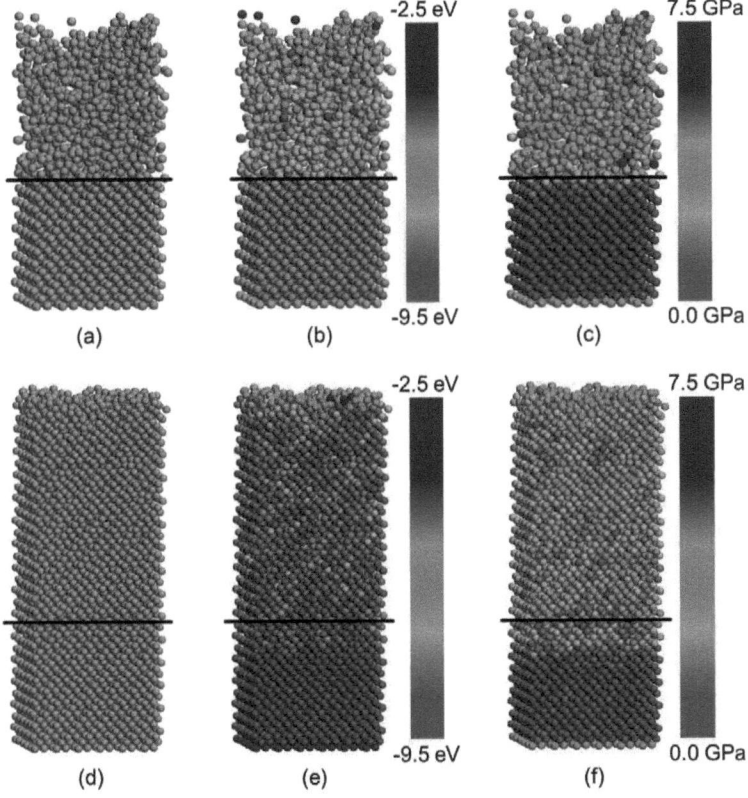

Figure 9.2: Silicon coatings grown on a Si (100) substrate. Top line: amorphous silicon coating (a) deposited at 500 K substrate temperature. The potential energy distribution is presented in (b), von Mises stress distribution in (c). In both cases one observes a clear substrate-coatings interface. Bottom line: high temperature deposition at 1000 K substrate temperature results in a crystalline silicon coating presented in (d).

Chapter 9. MD simulation of the deposition process

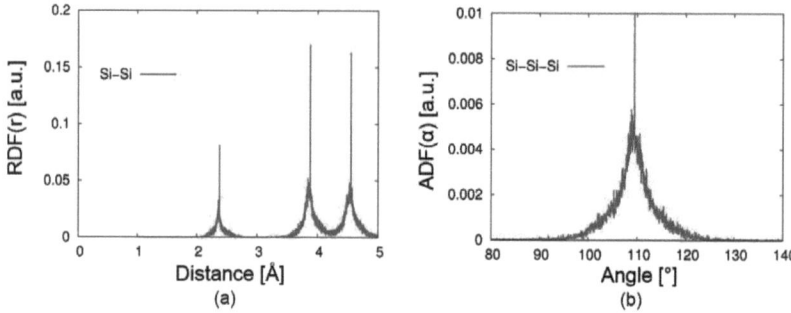

Figure 9.3: Radial (RDF) (a) and angular (ADF) (b) distribution function of the amorphous Si-substrate/Si-coating system as presented in **Figures 9.2(a)-9.2(c)**. Single peaks correspond to an ideal Si single crystal, while a broadening of both the radial and the angular distribution function can be observed due to the amorphous network of the growing Si coating.

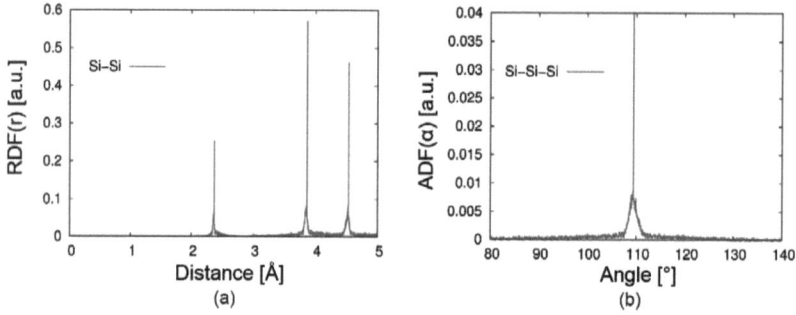

Figure 9.4: Radial RDF (a) and angular ADF (b) distribution function of the crystalline Si-substrate/Si-coating system as presented in **Figures 9.2(d)-9.2(f)**. Individual peaks correspond to single crystal Si (100) substrate. Smaller, triangular distributions at the bottom of the individual appear due to thermal oscillations within Si-coating.

Section 9.2. Deposition process of silicon on a Si substrate

9.2.2 Thermal annealing of silicon coatings

Dependent on the selected process paramaters, the structure of the resulting silicon coating varies, as presented in the previous section, ranging from loosely bonded, low-density silicon coatings as a result of the low temperature and/or low deposition energy process to crystalline silicon coatings with well defined material properties, which are normally grown at higher substrate temperatures.

In this section, an MD study of a possible subsequent temperature treatment of amorphous silicon coatings is presented. For this purpose, a system of an amorphous silicon coating, deposited onto a Si (100) substrate is chosen as an example. The deposition process was simulated at 500 K substrate temperature, a deposition rate of 0.1 particle/ps and the deposition energy of 1.5 eV, the resulting Si-substrate/Si-coating system is presented in **Figures 9.2(a)-9.2(c)**. Temperature controlled MD simulation of thermal annealing of such system is presented in this section. The annealing simulation can be divided into three stages: at first, the system temperature increases gradually from the initial 500 K to 1000 K within 1 ns. Secondly, a constant temperature of 1000 K is maintained for further 3 ns, and at last 1 ns cooling period to the initial 500 K is imposed. Time developement of the average temperature during the thermal annealing process is presented in **Figure 9.5(a)**. Global potential energy, averaged over all atoms witin the described system was monitored, its development is presented in **Figure 9.5(b)**. The initial potential energy of the system is higher than the average potential energy of bulk silicon at the same temperature. At first, the potential energy increases continuously for the first 1 ns, indicating the increasing particle movement caused by external heating. Once the system has reached 1000 K, the averaged potential energy decreases slowly, since the initial amorphous structure of the silicon coating rearranges, allowing atoms to obtain energetically more favoured positions of the silicon single crystal lattice. A subsequent simulation of cooling towards the initial 500 K further decreases the potential energy in the way that it becomes lower than at the beginning of the annealing simulation.

Chapter 9. MD simulation of the deposition process

Besides global variables, such as average temperature and the potential energy, presented in **Figure 9.5(a)** and **(b)**, local distributions of the potential and kinetic energy of the individual particles as well as of stresses present in the Si-substrate/Si-coating system can be analyzed. In **Figure 9.6** the system structure is presented, while following figures present the distribution of the potential energy in **Figure 9.7**, temperature in **Figure 9.8** and von Mises stresses in **Figure 9.9** within the system during the annealing simulation. In all cases **Figures (a)** present the initial system at 500 K, **Figures (b)** the system at 1000 K and the beginning of thermal annealing, **Figures (c)** the system at the end of thermal annealing process at 1000 K and **Figures (d)** the simulation at the final stage at 500 K.

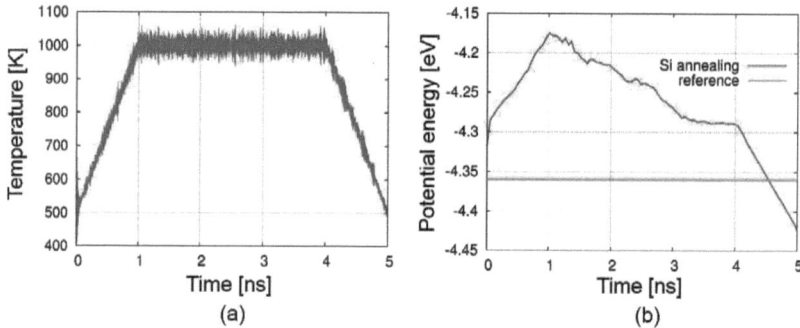

Figure 9.5: Development of the temperature and the averaged potential energy of the Si-substrate/Si-coating system (see **Figures 9.2(a)-9.2(c)**) during the annealing process. The green reference line at E_{pot}=-4.36 eV in **Figure (b)** indicates the average potential energy of a silicon single crystal of the same size as the presented Si-substrate/Si-coating system.

In addition to the analysis of the temperature, potential energy and stress distribution, radial (RDF) and angular distribution function (ADF) provide a good insight into the structures present in the Si-substrate/Si-coating system, before and after the thermal annealing process. **Fig-**

ures **9.3(a)** and **9.3(b)** show the predominant distances and angles in a such system, in both cases the peaks are located at the distance of the first, second and third nearest neighbor and the 109.5 ° angle respective of the silicon lattice. However, significant broadening of individual distributions can be observed, especially in direct comparisson with the high temperature deposition and the resulting crystalline silicon coating, RDF and ADF presented in **Figures 9.4(a)** and **9.4(b)**. **Figures 9.10(a)** and **9.10(b)** present the radial and the angular distribution function of the Si-substrate/a-Si-coating system after the described thermal annealing process. Pronounced peaks at the location of the first, second and third nearest neighbor of the silicon structure can be observed as well as the typical angle of 109.5 °. The broadening of individual distributions is not present any more, resembeling the the RDF and ADF of a crystalline Si-coating as presented in **Figures 9.4(a)** and **9.4(b)**.

Figures 9.5-9.10 present in a vivid manner that a subsequent thermal annealing of an amorphous silicon coating leads to its crystallization, hereby obtaining desired properties of a crystalline material. This can be achieved due to simple phase diagram of silicon, containing only one base atom and no stable intermediate phases. A counter-example is SiC, which has 2 atoms in the crystal structure base. In SiC, the metastable graphite phase and the formation of C-C dimers, significantly slows down the recrystalization process, making its modeling, at least on the MD scale of several ns, not possible.

It may not be always possible to deposit crystalline silicon thin films. This can for example be the case due to low thermal stability of the substrate material, when e. g. the substrate material has a lower melting point than the coating materials or the bulk modulus of the substrate material is to temperature sensitive. The described MD method of silicon annealing is simple and fast, allowing the exact analysis of local variables such as temperature, potential energy and residual stresses, however, a confirmation of the present results by numerical and/or experimental results of other research groups has to be achieved in order

Chapter 9. MD simulation of the deposition process

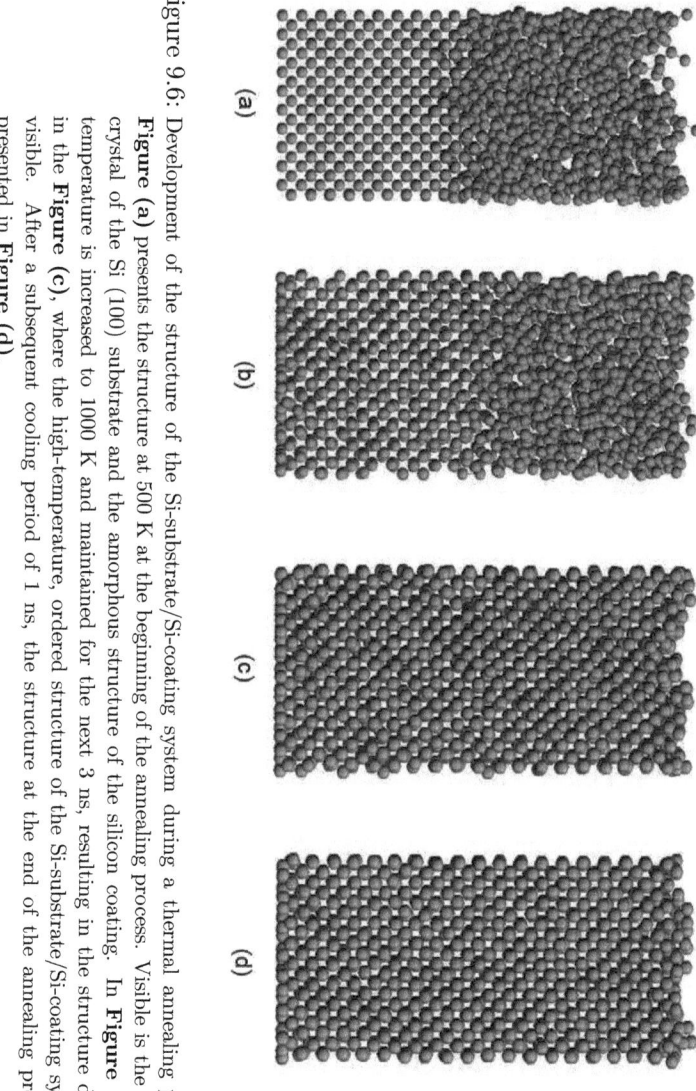

Figure 9.6: Development of the structure of the Si-substrate/Si-coating system during a thermal annealing process: **Figure (a)** presents the structure at 500 K at the beginning of the annealing process. Visible is the ordered crystal of the Si (100) substrate and the amorphous structure of the silicon coating. In **Figure (b)** the temperature is increased to 1000 K and maintained for the next 3 ns, resulting in the structure depicted in the **Figure (c)**, where the high-temperature, ordered structure of the Si-substrate/Si-coating system is visible. After a subsequent cooling period of 1 ns, the structure at the end of the annealing process is presented in **Figure (d)**.

Section 9.2. Deposition process of silicon on a Si substrate

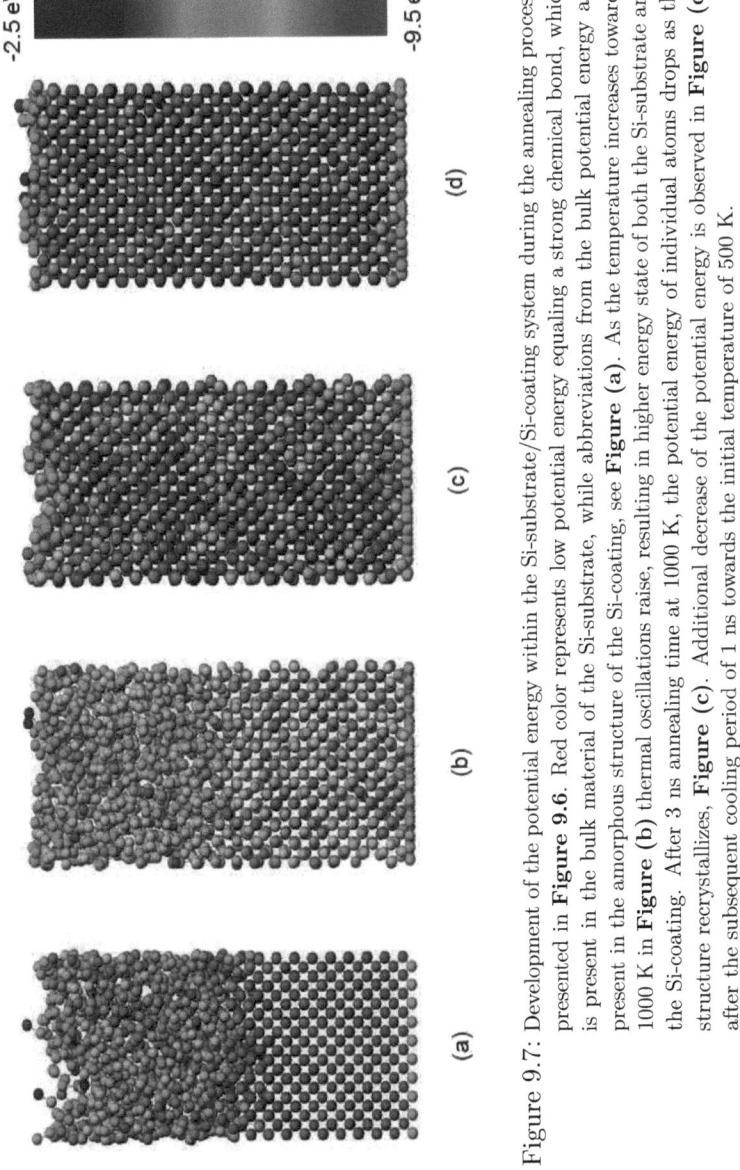

Figure 9.7: Development of the potential energy within the Si-substrate/Si-coating system during the annealing process, presented in **Figure 9.6**. Red color represents low potential energy equaling a strong chemical bond, which is present in the bulk material of the Si-substrate, while abbreviations from the bulk potential energy are present in the amorphous structure of the Si-coating, see **Figure (a)**. As the temperature increases towards 1000 K in **Figure (b)** thermal oscillations raise, resulting in higher energy state of both the Si-substrate and the Si-coating. After 3 ns annealing time at 1000 K, the potential energy of individual atoms drops as the structure recrystallizes, **Figure (c)**. Additional decrease of the potential energy is observed in **Figure (d)** after the subsequent cooling period of 1 ns towards the initial temperature of 500 K.

Chapter 9. MD simulation of the deposition process

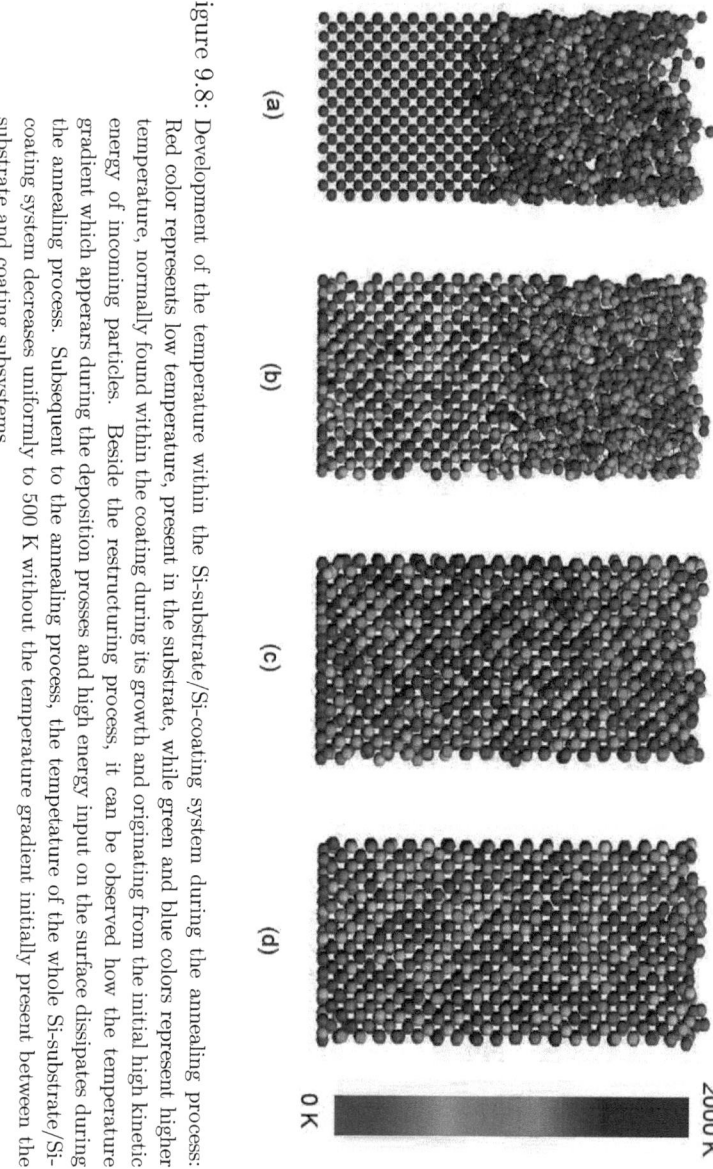

Figure 9.8: Development of the temperature within the Si-substrate/Si-coating system during the annealing process: Red color represents low temperature, present in the substrate, while green and blue colors represent higher temperature, normally found within the coating during its growth and originating from the initial high kinetic energy of incoming particles. Beside the restructuring process, it can be observed how the temperature gradient which appears during the deposition prosses and high energy input on the surface dissipates during the annealing process. Subsequent to the annealing process, the tempetature of the whole Si-substrate/Si-coating system decreases uniformly to 500 K without the temperature gradient initially present between the substrate and coating subsystems.

Section 9.2. Deposition process of silicon on a Si substrate

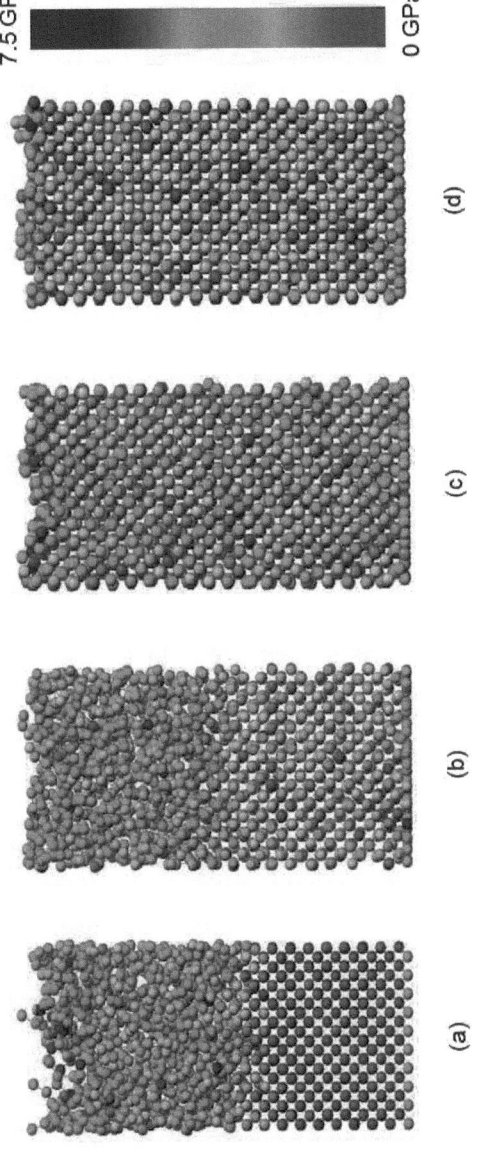

Figure 9.9: Development of von Mises equivalent stresses within the Si-substrate/Si-coating system during the annealing process: The same color representation is chosen as in **Figures 9.6-9.8**. Evident is the low stress state of the substrate, while higher stresses are present within the coating, see Figure (a). During the annealing process, residual stresses within the coating decrease gradually while the coating crystallizes.

Chapter 9. MD simulation of the deposition process

for our statements to gain significance. Humbird and Graves reported in [38] the molecular dynamics simulations of the sputtering and annealing process of silicon by Ar^+ ions using a Stillinger-Weber potential [48] and a Molière pair potential [144]. In **Chapter 7, Section 7.2** of this work, a good relation between the present simulation results and those obtained numerically by Humbird and Graves on the example of Ar^+ penetration depths in silicon has already been reported. From the aspect of annealing of silicon, there also exists a large amount of overlap with present simulation results, however, the thermal energy needed for the silicon structure annealing originates in the case of [38] from subsequent impacts of Ar^+ ions.

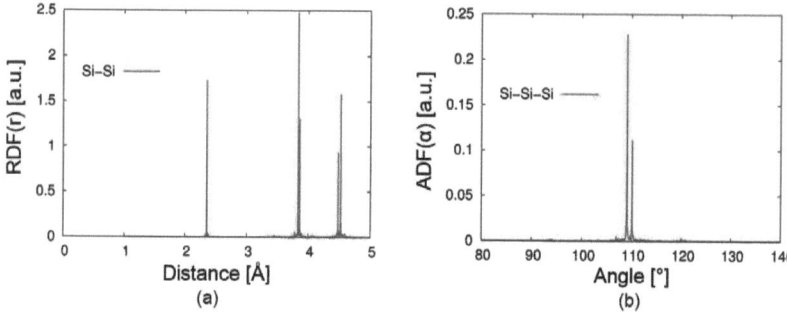

Figure 9.10: Radial (RDF) (a) and angular (ADF) (b) distribution function of the Si-substrate/Si-coating system after the annealing process presented in **Figures 9.5-9.9**. Both the RDF as well as ADF show a perfect crystal structure with well pronounced first, second and third nearest neighbours as well as the 109.5 ° angle of the diamond tetrahedra. The double peak in the ADF, **Figure (b)** is a concenquence of the small size of the system, containing only 4000 atoms with two free surfaces in ± z-direction and can, therefore, be ignored.

In [142] the authors report on molecular dynamics simulations of the recrystallization process of silicon, based on the movement of the amorphous/crystalline (a/c) interface by recombination processes. As pre-

Section 9.2. Deposition process of silicon on a Si substrate

sented here in **Figures 9.6-9.9** the recrystallisation of the amorphous Si-coating starts from the substrate-coating interface and propagates through the system, making the presented results in accordance with those in [142]. The simulation results in [142] are also supported by experimental high-resolution electron microscopy (HREM) investigations on silicon wafers, however, the results lack the analysis of local variables such as temperature, potential energy and residual stresses, which are presented here.

Besides other simulation results which were found to be supportive for present observations, a large amount of experimental research has been done on the field of silicon coatings and possible industrial application of the same. In 2000 Kim et al. [141] report about scanning rapid thermal annealing (RTA) process for polycrystalline silicon thin film transistors (TFTs). Amorphous silicon thin films were deposited on a glass substrate and later on annealed by tungsten-halogen lamps. For schematics of the experimental apparatus, the readers referred to Figure 1 in [141]. The linear light beam of the lamp was focussed using an elliptical reflector onto the a-Si and scanned over the specimen. The resulting thermal profile of the silicon specimen was controlled by both scan speed and lamp power. The described method enables the fabrication of low cost, polycrystalline silicon TFTs on conventional glass substrates within a few minutes, hereby making it an attractive method for industrial applications. Han et al. describe in [26] the thermal annealing process of amorphous silicon thin-film transistors deposited at 150 °C on a flexible stainless steel substrate, which represents a further development of the work published in [141]. In this publication from 2007, it was also reported that silicon on a flexible steel was the most promising candidate for future active matrix organic light-emitting diodes, or *AMOLEDs*, due to its strenght and high corrosion stability. Today, only a few years later, the abbreviation form *AMOLED* is widely known for its application in displays for mobile devices, confirming the foresight of the authors.

Chapter 9. MD simulation of the deposition process

9.3 SiC deposition process on a Si substrate

After having investigated the basic principles of the coating formation on the example of silicon, the material system was extended to the deposition process of SiC on silicon. Identical process parameters such as the substrate temperature, deposition rate and deposition energy were used and their influence on the coating structure was investigated.

For all remaining process parameters, the deposition process at higher deposition rate of 1 particle/ps resulted in amorphous SiC-coatings, indicating the deposition rate to be one of the most important process paramaters. At the lower deposition rate of 0.1 particle/ps the obtained SiC-coatings resulted in a more crystalline structure then for the deposition rate of 1 particle/ps. Next to the deposition rate, the influence of the substrate temperature has proven to be a critical paramater as well. Low substrate temperatures, up to 500 K resulted allways in amorphous coatings.

In general, a higher deposition energy had a positive effect on the formation of crystalline coatings. The impact of the deposition energy on the coating structure resembles that of the substrate temperature, in both cases, an increase of the average kinetic energy within the forming coating is found, resulting in an easyer surmounting of potential barriers caused by other atoms and favoring the formation of ordered structures.

At a lower deposition rate of 0.1 particle/ps, a substrate temperature of 1000 K and the deposition energy of 3 eV, it was possible to deposit crystalline SiC-coatings on a Si substrate for the first time without the use of fullerenes as precursor [51]. These observations were made both within MD simulations, presented in **Figure 9.11** as also in experiments, performed at KIT-IAM AWP and presented in **Figure 9.13**. **Figure 9.11** presents a comparison of stoichiometric SiC-coatings deposited at low (500 K) and high (1000 K) substrate temperatures.

Section 9.3. SiC deposition process on a Si substrate

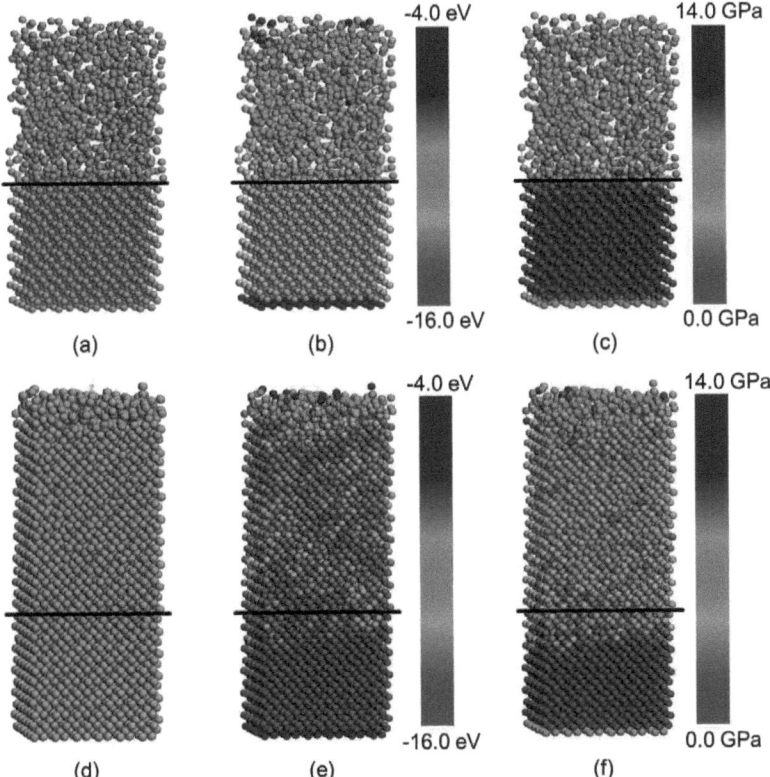

Figure 9.11: Stoichiometric SiC-coatings grown under different deposition conditions, e.g., temperature. Top line represents an amorphous SiC-coating deposited at 500 K Si (100) substrate temperature, 0.1 particle/ps deposition rate and 10 eV deposition energy. In (a) the system structure is presented, showing the Si (100) substrate and the amorphous SiC-coating on top. In (b) the distribution of the potential energy is presented, revealing reletively high potential energy (green color) of the amorphus SiC structure. In (c) the distribution of the von Mises stresses can be observed, identifying the low stress state of the silicon substrate (in red) and the high stress state of the SiC-coating (green color). In the bottom line high temperature deposition at 1000 K substrate temperature and the remaining parameters unchanged is shown resulting in a crystalline SiC-coating. One can notice the ordered structure of the coating in (d) and lower average potential energy within the coating (e). The stress analysis (f) reveals smoother transit from the low stress state of the substrate towards the SiC-coating.

Figure 9.11(a) makes it evident that a low-density, amorphous SiC-coating is formed directly upon the substrate at the lower substrate temperature of 500 K, while an ordered, crystal-like coating grows at higher substrate temperatures. **Figures 9.11(b)** and **9.11(e)** further reinforce this statement by presenting a higher average potential energy within the amorphous structure of the SiC-coating (visible through the yellow and green colored atoms). In the case of a crystalline SiC-coating, most atoms obtain their potential minimum, resulting in a predominantly red color in **Figure 9.11(e)**. Another important information is provided by the analysis of residual stresses. Due to the latice missmatch of approx. 20 % (5.5 Å silicon lattice constant and 4.3 Å for SiC), high stresses at the substrate-coating interface should be expected. Both the amorphous as well as the crystalline SiC-coating exibit these stresses on the Si-substrate (and vice versa) during the film growth. **Figures 9.11(c)** and **9.11(f)** present the distribution of von Mises stresses within the Si-substrate/SiC-coating system. From **Figure 9.11(c)** it is evident the silicon substrate shows a low stress state while the amorphous SiC-coating is exposed to higher von Mises stresses due to its structure. This picture changes somehow in the case of a crystalline SiC-coating. Again, the silicon structure shows a lower stress state, however, only in its lower atomic layers. The upper substrate layers, closer to the substrate-coating interface show higher stress states than it is the case for an amorphous SiC-coating, indicating that a crystalline SiC-coating exibits stronger influence onto the silicon substrate than its amorphous representative. As for the SiC-coating, it still shows high von Mises stresses (predominantly green color in **Figure 9.11(f)**), their origins need some attention.

Figure 9.12 presents the radial distribution function of the Si-substrate/c-SiC-coating system. Starting from the silicon substrate and building the SiC-coating layer by layer, the substrate structure remains dominant over the coating, hereby straining the coating structure to the silicon crystal lattice. At smaller coating thicknesses, this is the main reason for a high stress state within the coating as presented in

Section 9.3. SiC deposition process on a Si substrate

Figure 9.11(f).

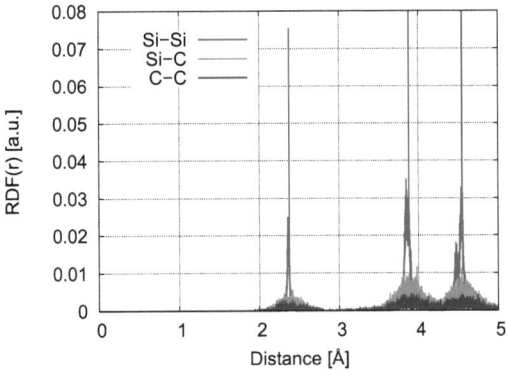

Figure 9.12: Radial distribution function (RDF) of a crystalline Si-substrate/SiC-coating system as presented in **Figures 9.11(d)-9.11(f)**. Noticable is that both Si-C and C-C bonds show the same lenght as Si-Si bonds indicating the dominance of the silicon substrate onto the SiC-coating. As a reference, the first, second and third nearest neighbour distances are $1^{st}=2.38$ Å, $2^{nd}=3.87$ Å and $3^{rd}=4.6$ Å in silicon, respectively $1^{st}=1.86$ Å, $2^{nd}=3.0$ Å and $3^{rd}=3.6$ Å in β-SiC.

The experimental analogon of these observations is presented in **Figures 9.13-9.16**. **Figure 9.13** is obtained by an atomic force microscopy (AFM) and shows the surface structure of a SiC-coating deposited at a silicon sbstrate at 600 °C and 700 °C. It is evident that in the case of a high temperature deposition crystalline SiC-coatings are grown with the average crystallite size of 8 nm. On the contrary, only amorphous SiC-coatings could be deposited at the lower substrate temperature of 600 °C.

Similar obserations are made by X-ray diffractometry (XRD) investigations. **Figure 9.14(a)** shows the dependence of the SiC-coating structure onto the substrate temperature. A first weak reflection is obtained for 700 °C, indicating the formation of SiC-crystallites. The initially small reflection intensity increases significantly for the further

Chapter 9. MD simulation of the deposition process

increase of the substrate temperature to 800 °C and 900 °C indicating increasing crystallite sizes. The SiC crystallite size as a function of silicon substrate temperature is presented in **Figure 9.14(b)**, showing that at the lower deposition temperature below 600 °C only amorphous SiC-coatings are grown, with a rapid increase of the crystallite size of deposited SiC-coatings above 700 °C substrate temperature.

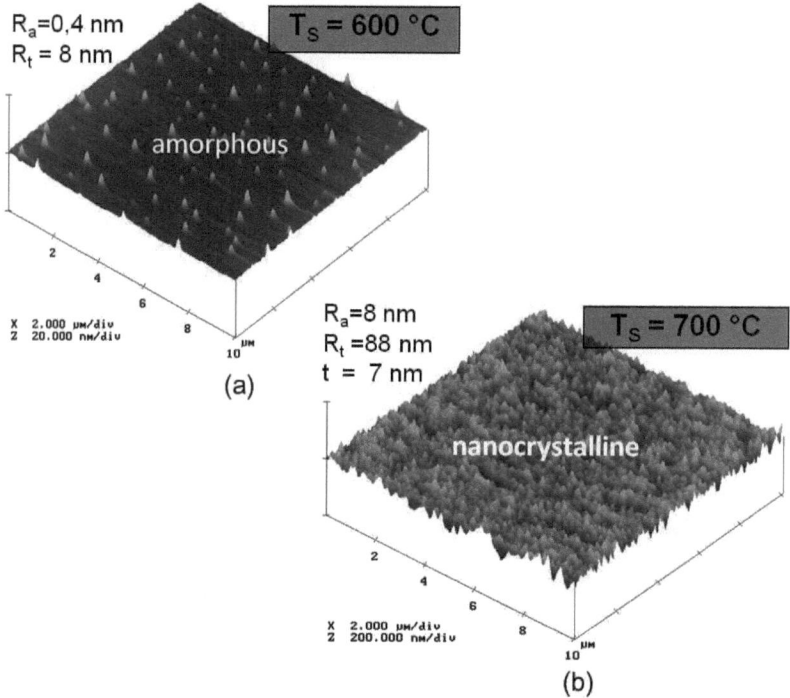

Figure 9.13: Comparison of the surface tomographies made by atomic force microscopy (AFM) of SiC-coatings deposited at a silicon substrate temperature of 600 °C in (a) and 700 °C in (b).

Section 9.3. SiC deposition process on a Si substrate

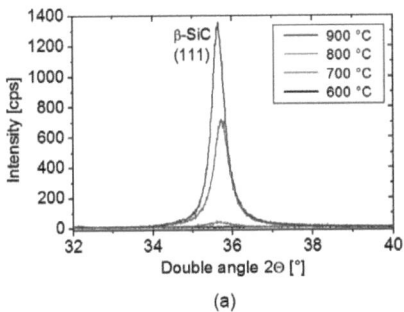
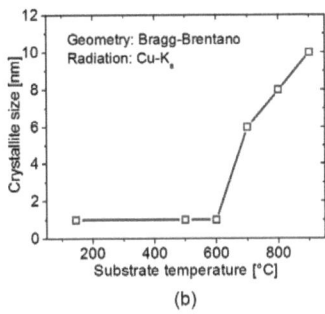

Figure 9.14: Experimental analysis of the temperature dependence of the SiC-coating structure. X-ray diffractogram (XRD) (a) of a SiC-coating at constant bias voltage $U_S = 0$ V in the temperature range 600 °C-900 °C and the crystallite size as a function of the substrate temperature (b).

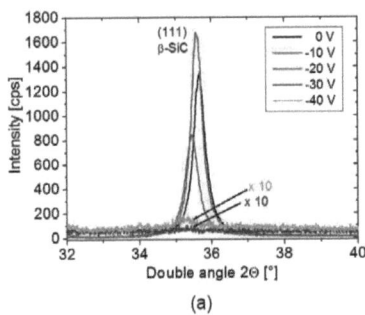
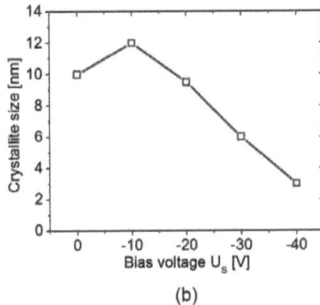

Figure 9.15: Experimental analysis of SiC-coating structure dependence of the bias voltage. X-ray diffractogram (a) of a SiC-coating at constant substrate temperature $T_S = 900$ °C in the bias voltage range 0 V to -40 V and the crystallite size dependence (b). The experimental analysis of the lattice constant of SiC-coatings shows the value similar to those of a SiC bulk material. This effect is known as *nanostabilization* and is contrary to the SiC-coating structure being strethed to the Si lattice constant as presented in **Figure 9.12**. The effect of nanostabilization can be observed at greater coating thicknesses from approx. 90 nm.

Chapter 9. MD simulation of the deposition process

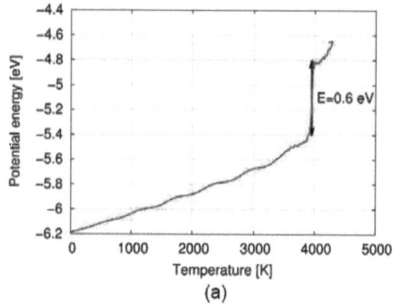
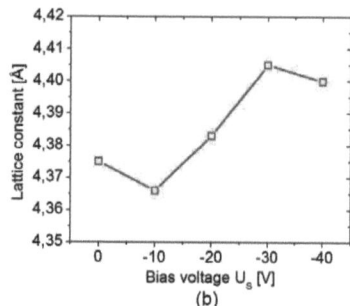

(a)   (b)

Figure 9.16: Molecular dynamics simulation of the melting of β-SiC, visible is the phase transition at 3900 °C (a). Lattice constant of a SiC-coating, deposited at 900 °C silicon substrate temperature, as a function of bias voltage (b).

Both **Figure 9.14(a)** and **Figure 9.14(b)** are obtained without bias voltage U_0, hence the kinetic energy of the film-forming particles originates only from the energy provided by impacting argon ions. Application of an additional bias voltage increases the kinetic energy of incoming particles and directly influences the coating formation process. **Figures 9.15(a)** and **9.15(b)** present the X-ray diffractogram and the crystallite size dependence as function of the bias voltage at constant substrate temperature of 900 °C. The X-ray diffractogram shows an intensity increase of approx. 20 % for $U_0 = -10$ V in comparison to the SiC deposition without the bias voltage. For the same case, the crystallite size also increases from 10 nm to 12 nm. Further increase of the bias voltage and hereby the kinetic energy of the incident particles diminishes the quality of the coating structure, the XRD-intensity decreases, so does also the crystallite size. Probable interpretation of this effect is that the coating formation is a competing process of the coating growth itself (favoured by increased deposition energy) and the coating destruction caused by back sputtering of particles forming the coating by those which are incoming. The latter process is favored by applying a bias voltages below $U_0 = -10$ V.

148

Section 9.3. SiC deposition process on a Si substrate

9.3.1 Silicon carbide gradient coatings

The deposition process of silicon by SiC was described in the previous section. The dependency of the coating structure on the deposition paramaters such as substrate temperature, deposition rate and deposition energy was analysed both by MD simulations and experimental investigations. High substrate temperatures and deposition energy lead to structured, crystalline coatings while an off-set of any process paramater leads to amorphous coatings. Regardless of the coating structure, high stresses have been observed on the substrate-coating interface. These stresses can lead to possible coating delamination during abrasive usage. It is, therefore, worth analyzing if the stresses on the substrate-coating interface can be reduced. In this section, the interface modification by a variation of the concentration of individual film-forming particles is presented, resulting in so-called gradient coatings.

Figure 9.17 presents the scheme of the approach. Rather than coating silicon directly by a stoichiometrical SiC-coating (Si:C particle ratio 1:1), the carbon concentration is increased over the coating thickness ranging from 0 % at the beginning to 50 % at the end. **Figure 9.17(a)** shows a silicon (100) substrate and the stoichiometrical SiC-coating on top of it, the substrate-coating interface with the 20 % lattice missmatch is shown in the red frame. In the case of **Figure 9.17(b)** the initial interface is divided into two interfaces with a smaller change of carbon concentrations, the silicon substrate is coated by $Si_{0.75}C_{0.25}$, in the second step a stoichiometrical SiC-coating is deposited. **Figure 9.17(c)** goes in the same direction, however, the concentration of carbon atoms is increased more slowly, resulting into four instead of two interfaces, with smaller steps regarding coating stoichiometry. In **Figure 9.17(b)** and **9.17(c)**, red arrows indicate the location of the interface, the arrow thickness indicates its intensity. The second line in **Figure 9.17** presents the result of MD simulations of the described depositions, silicon atoms are colored in yellow while carbon atoms are colored in gray. Visible is the increase of carbon concentration over the coating thickness in **Figures 9.17(d)-9.17(f)**.

Chapter 9. MD simulation of the deposition process

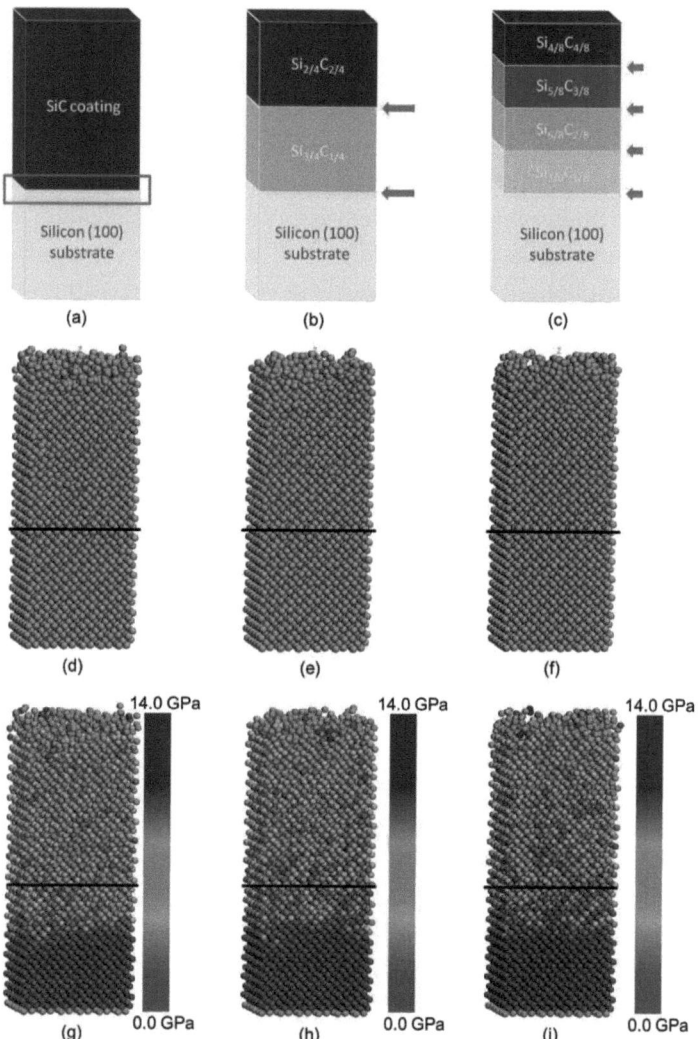

Figure 9.17: SiC gradient coatings on a Si (100) substrate: schematics of different coating stoichiometries (top line) with red arrows indicating individual interfaces. The particle distributions are presented in the second line (silicon atoms in yellow, carbon atoms in gray) while the resulting von Mises stress distribution is shown in the bottom line (low stress in red, high stress in green and blue).

The most interesting aspect of these simulations is surely the stress representation. **Figures 9.17(g)-9.17(i)** present the distribution of von Misses stresses within the Si-substrate/SiC-coating system. While in **Figure 9.17(g)** a clean interface is shown (color change red to green), this interface is dissolved into a region of different widths, depending on the carbon concentration. The stress decrease is evident in the case of **Figure 9.17(h)** and **9.17(i)**.

Figure 9.16(a) shows an MD simulation of the melting of β-SiC. As it can be observed, the calculated melting point lies around 3950 K, this is an overestimation of the experimentally measured melting temperature of SiC at 2700 °C. The overestimation of the melting point is common in molecular dynamics simulations, since the majority of interatomic potentials is fitted to describe the solid state a a material under investigation. The important value of **Figure 9.16(a)** is the phase transition energy of 0.6 eV, which differs the crystalline and the amorphous solid state. Going from the right hand side of **Figure 9.16(a)** to the left, the formation of a crystalline SiC-coating is energetically favoured process. This process is, however, hindered by the Si/SiC lattice mismatch of approx. 20 %.

9.4 Si$_3$N$_4$ deposition process on a Si substrate

Modelling the deposition process of silicon by Si$_3$N$_4$ was performed in the same way as for the example of SiC on silicon. **Figure 9.18** presents the result of the deposition process of stoichiometric Si$_3$N$_4$-coatings as function of substrate temperature. In **Figures 9.18(a)-9.18(c)**, a low temperature deposition at 500 K is presented. The deposition process resulted in amorphous Si$_3$N$_4$-coatings, regardless of the remaining deposition parameters. For high temperature deposition at 1000 K, low deposition rate of 0.1 particle/ps and deposition energy of 3 eV or 6 eV, stoichiometric Si$_3$N$_4$-coatings showed ordered, crystal-like structures. The results are presented in **Figures 9.18(d)-9.18(f)**. Again, as in the case of SiC-coatings, a higher potential energy within the coating is

Chapter 9. MD simulation of the deposition process

characteristic for amorphous Si_3N_4-coatings (see **Figures 9.18(b)** and **9.18(e)**).

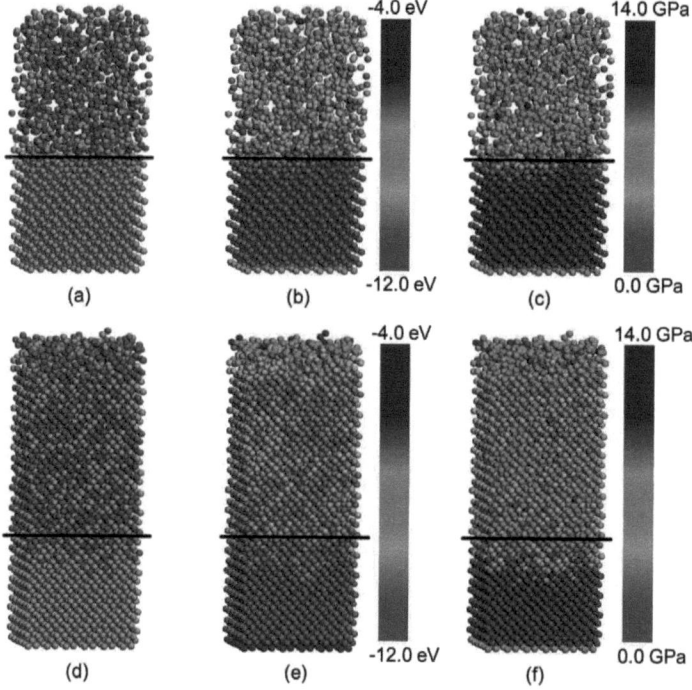

Figure 9.18: Stoichiometric Si_3N_4-coatings grown under different deposition conditions e. g. temperature. The top row represents an amorphous Si_3N_4-coating deposited at 500 K substrate temperature (low temperature deposition) while in the bottom row high temperature deposition at 1000 K substrate temperature is presented, resulting in a crystalline Si_3N_4-coating. The amorphous structure of (a) as well as the clean transition between the low stress state substrate and high stress state coating (c) can be observed. In the case of a high temperature deposition process and the resulting crystalline Si_3N_4-coating, the stress transition between the substrate and the coating material is smoother indicating higher cohesion on the substrate-coating interface.

Section 9.4. Si₃N₄ deposition process on a Si substrate

Another property of amorphous Si$_3$N$_4$-coatings is a clean transition from the low stress state substrate to the high stress state coating material (**Figure 9.18(c)**) indicating a weak adhesion on the substrate-coating interface. In the case of crystalline Si$_3$N$_4$-coatings, both the substrate and the coating are strained on the interface to accomodate each other, resulting in an interlocking and smoother stress transition at the interface, **Figure 9.18(f)**. Since the obtained coating thicknesses are still small (< 10 nm), no transition between the silicon and Si$_3$N$_4$ structure could be observed, deposited atoms occupy silicon lattice sites, as shown by the analysis of the radial distribution function in **Figure 9.19**.

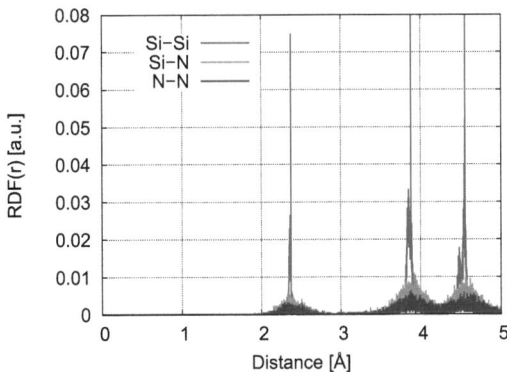

Figure 9.19: Radial distribution function of a crystalline Si-substrate/Si$_3$N$_4$-coating system as presented in **Figure 9.18**. Noticable is that both Si-N and N-N bonds show the same lenght as Si-Si bonds indicating the dominance of the silicon substrate onto the Si$_3$N$_4$-coating. The same observations were made for SiC-coatings of the same thickness, **Figure 9.12**.

9.4.1 Silicon nitride gradient coatings

A similar approach as in the case of SiC gradient coatings, **Section 9.3.1**, is applied to produce graded SiN-coatings.

Chapter 9. MD simulation of the deposition process

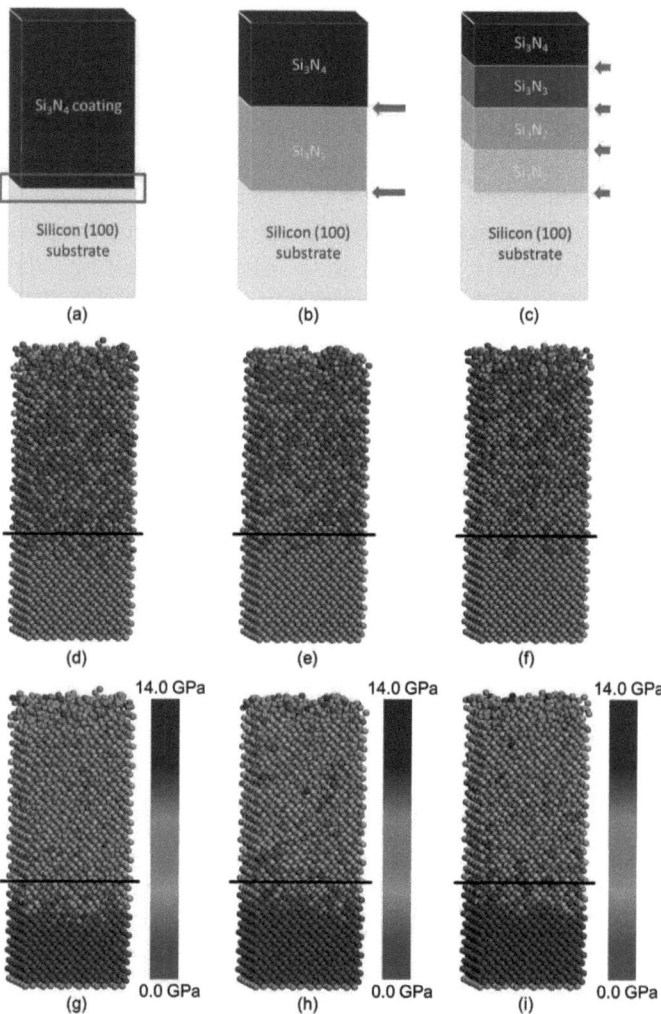

Figure 9.20: SiN gradient coatings on a Si (100) substrate: schematic comparison of different stoichiometries (top row), with red arrows indicating individual interfaces. The resulting particle distributions is given in the second row (silicon atoms in yellow, nitrogen atoms in blue) while the resulting von Mises stress distributions can be observed in the bottom row (low stress in red, high stress in green and blue).

Section 9.4. Si_3N_4 deposition process on a Si substrate

Due to different stoichiometry of Si_3N_4, it was possible to implement this approach by grading the nitrogen concentration only. **Figure 9.20** presents the results of the corresponding MD simulations. The left column presents a stoichiometric Si_3N_4-coating onto a silicon (100) substrate, while in the two remaining columns the nitrogen concentration was increased over the coating thickness in two (middle Figures Si_3N_2/Si_3N_4) or four steps (right Figures $Si_3N_1/Si_3N_2/Si_3N_3/Si_3N_4$). From **Figures 9.20(a)-9.20(i)** it becomes evident that a minor decrease in von Mises stress distribution is obtained by this approach. However, the results are not as significant as in the case of SiC gradient coatings. This may be the result of different, non-cubic structure of both α- and β-Si_3N_4 as well as as of non-trivial modelling of the nitrogen interaction in Si_3N_4. Another possible explanation for the absence of the desired result is that greather film thicknesses are needed for its observation. However, due to lacking of any experimental data in this topic, further assumptions should be avoided.

9.4.2 Experimental observations

Besides MD simulations, experimental investigations of the deposition process of Si_3N_4 on silicon were performed. **Figure 9.21(a)** presents an X-ray diffractogram of an α-Si_3N_4 single crystal target material. Individual peaks correspond to different crystal planes within the material. **Figure 9.21(b)** also shows an X-ray diffractogram, however, not an α-Si_3N_4 single crystal, but a stoichiometric Si_3N_4-coating on a silicon substrate is investigated. It can be observed that no individual peaks appear, indicating an amorphous structure. The high peak value at $2\Theta=70$ ° originates from the silicon substrate.

It is evident that in the case of Si_3N_4 deposition process MD simulations and experimental results contradict each other. While it was possible to obtain thin, crystalline coatings of Si_3N_4 in a numerical simulation for high substrate temperatures, an equivalent experimental observation has not been made. Possible explanations are small systems and thin

Chapter 9. MD simulation of the deposition process

coatings in MD simulations. On the simulation side, an interatomic potential poses also a certain source of errors due to relatively high ionicity of the Si-N bond, see **Figure 4.6** in **Chapter 4, Section 4.7**, which is not taken into account in the Tersoff potential form. Experimentally, an increase of the substrate temperature could lead to more ordered structures as in the case of SiC-coatings on silicon substrates, see **Figures 9.13-9.14**. In the framework of these investigations, silicon substrate temperatures up to 900 °C could be obtained. This leaves some space for further temperature increase since silicon melting point is given at $T_{melt}(Si)=1414$ °C. However, high substrate temperature combined with a high deposition energy and/or high deposition rate could lead to the melting of higher substrate layers and therefore negatively influence the coating growth.

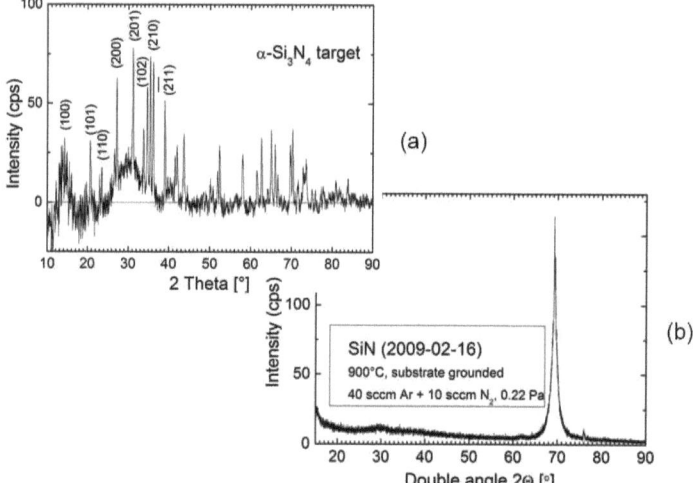

Figure 9.21: X-ray diffractogram of a crystalline α-Si$_3$N$_4$ target material (a) and an amorphous Si$_3$N$_4$-coating on a silicon substrate. Individual intensity peaks in (a) correspond to different crystallographic planes of the trigonal α-Si$_3$N$_4$-system. In (b) a peak at $2\Theta=70$ ° originates from the silicon substrate while the coating itself is X-ray amorphous.

10 Nanoindentation of Si, β-SiC and α-Si$_3$N$_4$ single crystals

In this chapter, molecular dynamics simulations of the nanoindentation process of Si, β-SiC and α-Si$_3$N$_4$ are presented. Nanoindentation of these materials was simulated by MD in order to gain a better insight into processes occuring at the atomic scale. Herefore, a rigid indenter formed as a Berkovich tip, **Figure 6.2** in **Chapter 6**, was used. The internal structure of the indenter was chosen to resemble the diamond crystal lattice (a=3.5 Å). The main intention of this approach was come close to the experimental situation, where most commonly diamond is used for indenter tips, hereby ensuring a high amount of comparability between the simulation and experiment.

Particle interactions within the substrate material were represented by the bond-order Tersoff potential [45, 74, 75, 76, 77, 78]. The usage of a bond-order potential is consistent with all previous simulations within this work and with Lin et al. [154]. In [154], however, only a nanoindentation of the Si (100) surface was presented, while the remaining silicon surfaces were not investigated. It is contrary to previous works of the group around Kalia and Vashishta, where a modified Stillinger-Weber potential [48] was used, with an explicit separation into two- and three-body terms. For the indenter-substrate interaction, a self-developed repulsive pair potential, described in **Section 4.6.1** and presented in **Figure 4.2** was implemented. The indenter tip was moved by a constant distance increment of 0.2 Å, a micro-convergence integrator MIK was applied to the substrate, representing a nanoindentation process at 0 K. Both the indenter-substrate potential as well as the indenter propagation increment were optimized to systems under investigation; nanoindentation of hard ceramics, such as SiC and Si$_3$N$_4$, leads to huge changes in interatomic forces even for small shifts in atomic coordinates. In addition, the cut-off radius of the indenter-substrate pair potentials was set very small, representing a hard sphere propagation and minimizing the energy input due to the indentation process. The approach of other research groups, which used a simplified version of available standard potentials, such as the repulsive part of the Tersoff potential [16], Morse potential [122, 154] or the ZBL-potential [84] (initially developed for high energy atomic collisions), was not adopted. In the authors opinion, the implementation of stated potentials into a nanoindentation simulation could compromise results, due to the fact that their development was motivated by other physical processes. During the MD simulation, atomic coordinates, forces, velocities, stresses and potential energies were calculated, allowing a comprehensive analysis of the nanoindentation process.

10.1 Application of the method by Oliver and Pharr

Mechanical properties, such as the hardness and Young's modulus (elastic modulus) were calculated according to the approach of Oliver and Pharr [151]. **Figure 10.1** shows a schematic representation of the indenter load-displacement relation for one loading and unloading cycle, where the parameter P designates the load and parameter h the displacement of the indenter tip in relation to the initially undeformed surface. From the P-h relation, three important quantities can be measured, the maximum indenter load P_{max}, together with the maximum indenter displacement h_{max} as well as the elastic loading stiffnes dS/dh defined as the slope of the upper part of the unloading curve. The accu-

Section 10.1. Application of the method by Oliver and Pharr

racy of hardness and modulus measurement depends inherently on how well these parameters can be measured experimentally [151], in an MD simulation, this problem becomes trivial since all atomic coordinates are well known at every stage of the simulation. For the calculation of the material hardness and the elastic modulus, not the maximum indentation depth h_{max}, but the contact depth h_c is used. The definition of h_c is evident from **Figure 10.2**, in the load-displacement relation, **Figure 10.1**, h_c is the value where the tangent of the upper portion of the unloading curve intersects the X-axis. h_c is smaller than the actual indentation depth due to the surface tension of the substrate material.

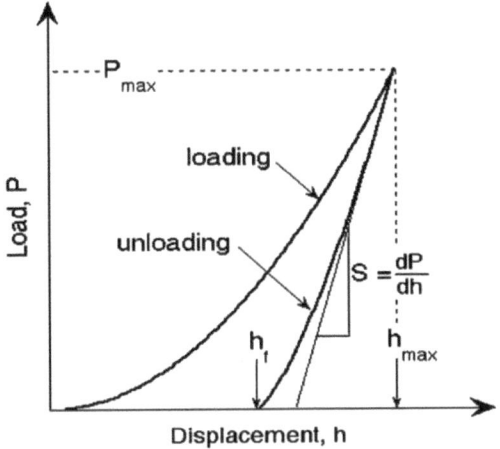

Figure 10.1: Schematic representation of the load-displacement relation during the nanoindentation experiment, source [151].

Having the maximum load P_{max} and the contact depth h_c, material hardness can be calculated as the ratio of the indenter load P_{max} and the projected area $A(h_c)$ in agreement with the equation:

$$H = \frac{P_{max}}{A(h_c)} \tag{10.1}$$

Chapter 10. Nanoindentation of Si, β-SiC and α-Si$_3$N$_4$ single crystals

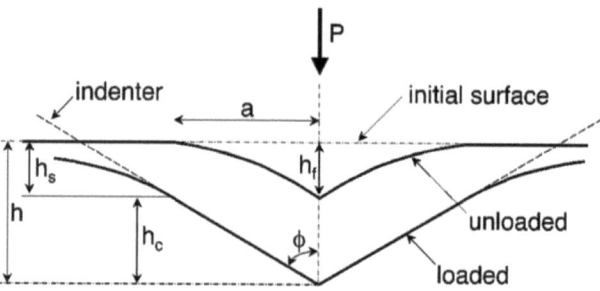

Figure 10.2: Schematic illustration of the unloading process showing parameters characterizing the contact geometry, source [151].

The projected area $A(h_c)$ depends on the geometry of the indenter tip, in the case of the three sided Berkovich indenter with an opening angle $\alpha = 143$ °, the projected area $A(h_c)$ can be calculated according to the equation:

$$A(h_c) = 24.49 \cdot h_c^2 \tag{10.2}$$

Measurement of the effective elastic modulus follows from its relationship to the contact area and the measured unloading stiffness through the equation:

$$S = \beta \frac{2}{\sqrt{\pi}} E_{\text{eff}} \sqrt{A(h_c)} \tag{10.3}$$

The parameter β presents the deviation of the indenter tip from an axisymmetric body, as sometimes present in experimental investigations. In the case of an MD simulation, a perfect form of the Berkovich indenter tip is present, therefore $\beta = 1$.

Hence, it can be stated that the effective elastic modulus E_{eff} can be represented as:

$$E_{\text{eff}} = \frac{S\sqrt{\pi}}{2\sqrt{A(h_c)}} \qquad (10.4)$$

In an experiment, the effective elastic modulus E_{eff} depends on the the elastic moduli and Poissons ratios of the specimen and the indenter material in agreement with the equation:

$$\frac{1}{E_{\text{eff}}} = \frac{1-\nu^2}{E} + \frac{1-\nu_i^2}{E_i} \qquad (10.5)$$

In the present MD simulations, the indenter tip is approximated as a rigid body with $E_i = \infty$, setting the second term of the equation 10.5 to zero. In addition, the MD simulation of nanoindentation is performed using fixed system sizes and therefore $\nu = 0$. Hence, an MD simulation of an nanoindentation using a rigid indenter tip allows the calculation of the real elastic modulus of the investigated material system:

$$E = E_{\text{eff}} = \frac{S\sqrt{\pi}}{2\sqrt{A(h_c)}} \qquad (10.6)$$

10.2 Nanoindentation of silicon

For the simulation of the nanoindentation process of silicon, a cubic Si system with a side length of approx. 30 nm and containing 1 000 000 particles was used. The lattice parameter was set to a=5.5 Å, as in the case of silicon at T=0 K. Three different silicon surfaces were nanoindented, the propagation direction of the indenter tip was parallel to one of the three low-index crystallographic axes, namely {100}, {110} and {111}. This was done in order to investigate the hardness dependence on the crystal orientation.

Chapter 10. Nanoindentation of Si, β-SiC and α-Si$_3$N$_4$ single crystals

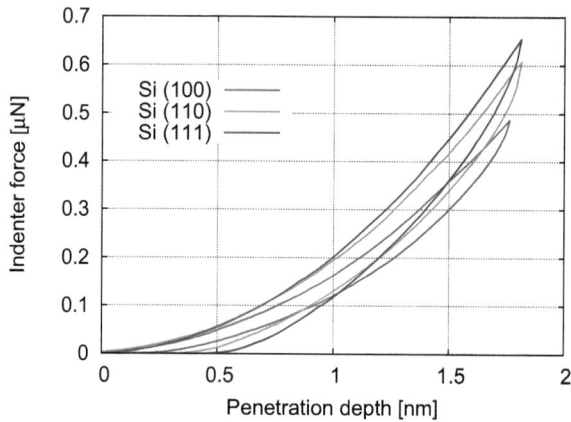

Figure 10.3: Indentation depth vs. indenter load for 3 low-index crystal orientatons of silicon.

Figure 10.3 presents the load-indentation depth relation of the Berkovich indenter tip during one loading and unloading cycle. It is evident that the load on the indenter tip increases considerably with the increasing indentation depth due to the large opening angle of the Berkovich indenter tip. All three crystal orientations show similar response, Si {100} shows the indenter load at the same indentation depth of 1.7 nm, being around 75 % of the indenter load of the Si {111} crystal orientation. For all Si orientations, the loading and unloading curve are very close to each other, indicating that a high amount material response is elastic.

According to Oliver and Pharr [151], the indenter load-indentation depth relation, **Figure 10.3**, allows the calculation of the material H hardness and elastic modulus E for individual cystal orientations of silicon. The calculated values are presented in **Table 10.1** and are in good relation to those found in [5] as well as own experimental investigations presented in **Figure 11.9** and [27].

Next to global data, such as the indenter load-indentation depth re-

Section 10.2. Nanoindentation of silicon

Orientation	Si {100}	Si {110}	Si {111}
Hardness [GPa]	7.8	8.3	9.4
E-Modulus [GPa]	178	270	224

Table 10.1: Silicon hardnesses and elastic moduli as calculated by molecular dynamics simulations and in dependence on the crystal orientation.

lation, the analysis of different local variables such as temperature, potential energy and stress tensor is important for the understanding of underlying atomic processes. Hence, distributions of all stated variables are analyzed and are presented in following figures.

Figure 10.4 presents the analysis of the potential energy distribution within the Si (100) single crystal at the maximum simulated indentation depth of the Berkovich indenter tip. For visualisation purposes, the indenter tip itself is not depicted, revealing the exact situation below. **Figure 10.4(c)** presents the top view on the system, while **Figures 10.4(a)** and **10.4(b)** show X- and Y-cross-sections through the silicon system at the location of the indenter tip. The potential energy distribution is presented in terms of different colors, red colored atoms present low potential energy, corresponding to non-disturbed crystal lattice, while, going over orange, yellow, green to blue, the whole range of potential energies within the specific simulation stage is depicted.

Top view, presented in **Figure 10.4(a)** reveals the three-sided imprint of the Berkovich indenter tip. **Figure 10.4** shows a very narrow distribution of the potential energy, localized around the indenter tip. From the point of view of the potential energy, the chosen system size was adequate, since all deviations from the average potential energy decay at the system boundaries. While the distribution of the potential energy is symmetric in **Figure 10.4(a)**, it is slightly asymmetric in the Y-cross-section, **Figure 10.4(b)**, this is due to the three-sided form of the Berkovich indenter tip. Dark blue atoms on the silicon surface present two-fold bonded, unsaturated silicon surface atoms, their potential energy is half of the potential energy inside the silicon bulk.

Next to the potential energy, the temperature distribution under the

Chapter 10. Nanoindentation of Si, β-SiC and α-Si$_3$N$_4$ single crystals

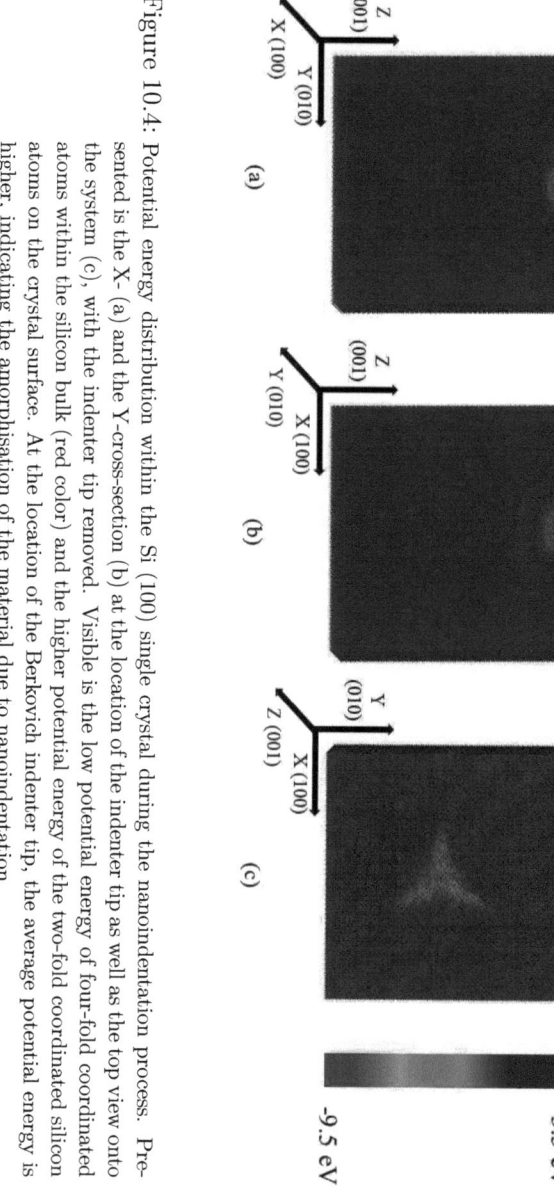

Figure 10.4: Potential energy distribution within the Si (100) single crystal during the nanoindentation process. Presented is the X- (a) and the Y-cross-section (b) at the location of the indenter tip as well as the top view onto the system (c), with the indenter tip removed. Visible is the low potential energy of four-fold coordinated atoms within the silicon bulk (red color) and the higher potential energy of the two-fold coordinated silicon atoms on the crystal surface. At the location of the Berkovich indenter tip, the average potential energy is higher, indicating the amorphisation of the material due to nanoindentation.

indenter tip is of great importance. Globally, a quasi-static state is assumed, corresponding to temperature of 0 K, however this does not exclude a local temperature deviation of an arbitrary amount. Special emphasis is, therefore, raised at this aspect of the simulation, since a large temperature deviation at a local level would, in the extreme case, lead to the melting of the system directly beneath the indenter tip, compromising results. In **Figure 10.5** the temperature distribution within the silicon sample is presented directly after the discrete indenter propagation of 0.2 Å. The presentation form is the same as in **Figure 10.4**, with the top view (c) and the X- and Y-cross-sections (a, b) at the location of the indenter tip. A small amount of yellow colored atoms in the vicinity of the indenter tip indicates a small, local temperature increase during the indentation process. This can be accounted to the carefull choice of the short-range pair potential describing the indenter-sample interaction, **Section 4.6.1**, which is often the object of criticism in the literature.

Next to load-penetration depth relation, the most important result of the nanoindentation simulation is the stress analysis. Hence, a distribution of individual stress components, as well as of the resulting von Mises stress is presented. **Figures 10.6(a)** to **10.6(c)** depict the top view representation of individual normal stresses. In all cases, smaller regions around the indenter tip differ from the global stress state. In the case of P_{xx} (**Figure 10.6(a)**) there is a stress increase corresponding to the location of the right plane of the Berkovich indenter tip, a local pressure distribution in the X-direction can very well be observed despite of the large opening angle of the indenter tip. The same statement is strengthened by the analysis of the P_{yy}-stress distribution (**Figure 10.6(b)**), a large stress increase is observed on the location of the left edge of the indenter tip, representing the stress increase due to cutting through the silicon surface. Evaluation of P_{zz} (**Figure10.6(c)**) reveals a similar picture, atoms directly unter the indenter tip are subjected to high stresses in the negative Z-direction, decreasing rapidly with increasing distance from the indenter tip.

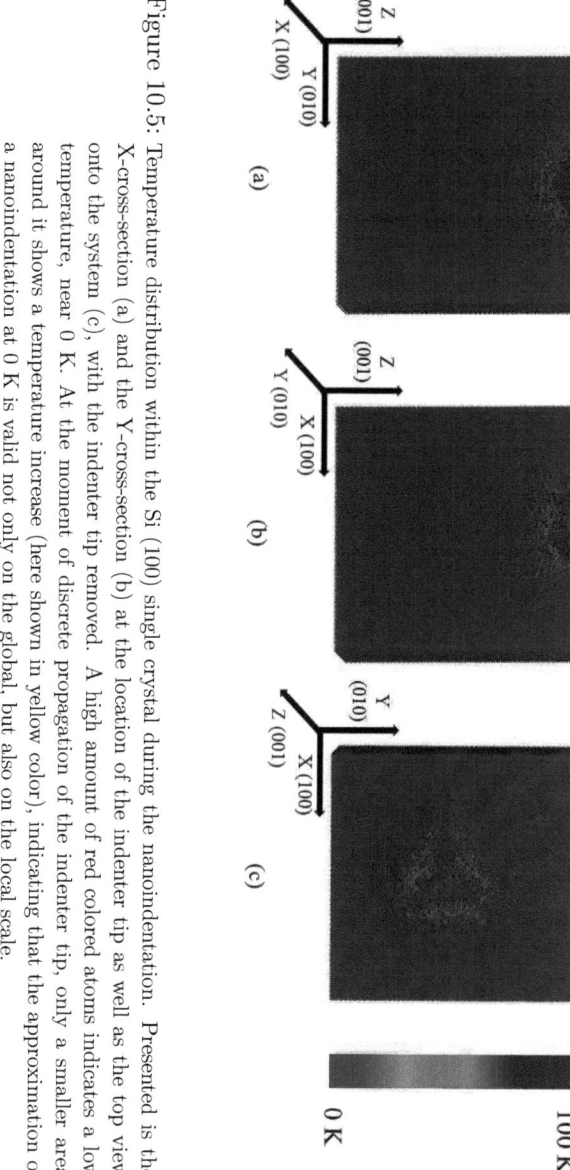

Figure 10.5: Temperature distribution within the Si (100) single crystal during the nanoindentation. Presented is the X-cross-section (a) and the Y-cross-section (b) at the location of the indenter tip as well as the top view onto the system (c), with the indenter tip removed. A high amount of red colored atoms indicates a low temperature, near 0 K. At the moment of discrete propagation of the indenter tip, only a smaller area around it shows a temperature increase (here shown in yellow color), indicating that the approximation of a nanoindentation at 0 K is valid not only on the global, but also on the local scale.

Section 10.2. Nanoindentation of silicon

Figure 10.7 also reveals an interesting distribution of shear stresses under the indenter tip. **Figure 10.7(a)** presents the P_{yz}-stress distribution; on the left side of the indenter tip two triangular regions are visible, corresponding to two sides of the Berkovich indenter tip. The two regions show different colors which represent opposite directions of the P_{yz}-shear stress. The remaining side of the indenter tip is visible in the **Figure 10.7(b)** and depicts the increased amount of P_{zx}-stress corresponding to the remaining plane of the Berkovich indenter tip.

Finally, von Mises stress distribution is presented in **Figure 10.8**. Again, **Figure 10.12(c)** presents the top view onto the system without depicting the indenter itself while two lower figures depict the X- (**Figure 10.8(b)**) and Y-cross-section (**Figure 10.8(c)**) at the location of the lowest indenter atom. This is probably the best representation of the local stress state of the system. Not only the radial extend of the stress distribution within the system can be observed, but also the symmetry (in the X-cross-section) and the asymmetry (in the Y-cross-section) imposed by the shape of the Berkovich indenter tip. A carefull reader will observe two additional features: first, the high stress state of the silicon surface (green atoms) caused by the tension during the indenter penetration and secondly the high stress state of bottom layer atoms; positions of these atoms were held fixed during the nanoindentation simulation, presenting the substrate holder in an experimental setup.

Next to the analysis of potential energies and stresses, deviations from a perfect crystal structure require special attention. It is well known that ductile materials such as fcc and bcc metals react to nanoindentation by the formation of dislocations. On the other hand, in the case of brittle materials such as ceramics, the dislocation formation and their propagation is energetically unfavoured, nanoindentation causes an amorphisation beneath the indenter tip, as described in [113] for TiC and VC. **Figure 10.9** presents the nanoindentation induced amorphisation of silicon. It is evident that the amorphous layer of silicon material is very thin, following the form of the Berkovich tip closely. Similar ob-

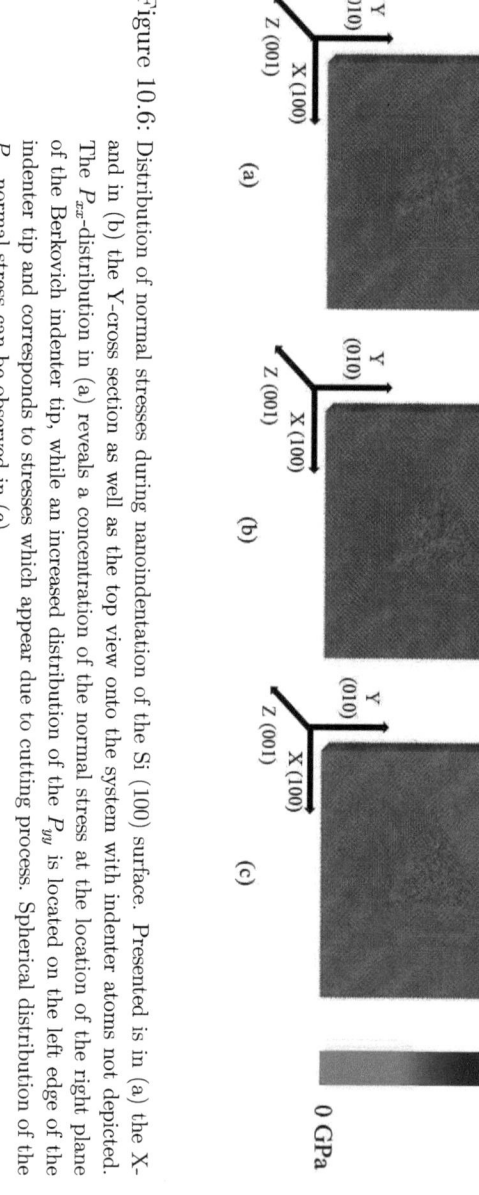

Figure 10.6: Distribution of normal stresses during nanoindentation of the Si (100) surface. Presented is in (a) the X- and in (b) the Y-cross section as well as the top view onto the system with indenter atoms not depicted. The P_{xx}-distribution in (a) reveals a concentration of the normal stress at the location of the right plane of the Berkovich indenter tip, while an increased distribution of the P_{yy} is located on the left edge of the indenter tip and corresponds to stresses which appear due to cutting process. Spherical distribution of the P_{zz} normal stress can be observed in (c).

Section 10.2. Nanoindentation of silicon

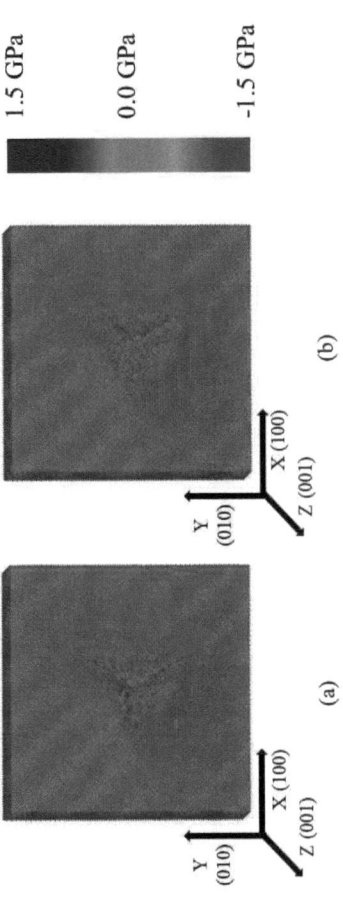

Figure 10.7: Distribution of shear stresses during the nanoindentation of the Si (100) surface. Presented are top views onto the silicon system with indenter atoms not depicted. Figure (a) shows the distribution of P_{yz}-shear stresses which are concentrated at the left edge of the Berkovich indenter tip. The opposite direction of P_{yz}-shear stresses on the two sides of the Berkovich indenter tip are visible trough the two triangle areas (red and blue color in figure (a)). A high concentration of P_{zx}-shear stresses can be observed in (b) at the location of the right plane of the Berkovich indenter tip.

Chapter 10. Nanoindentation of Si, β-SiC and α-Si$_3$N$_4$ single crystals

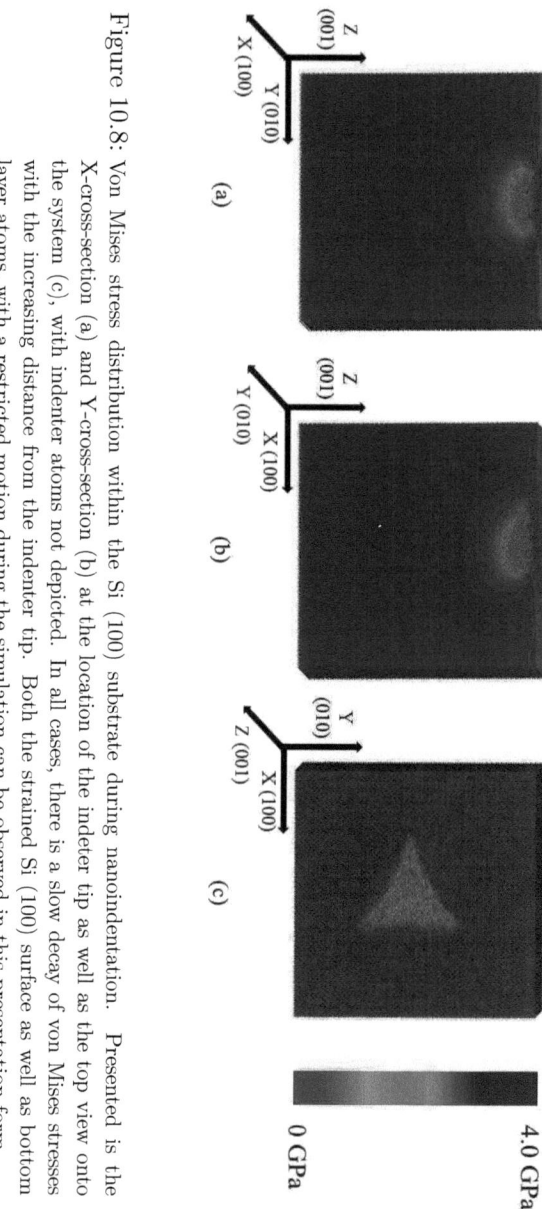

Figure 10.8: Von Mises stress distribution within the Si (100) substrate during nanoindentation. Presented is the X-cross-section (a) and Y-cross-section (b) at the location of the indenter tip as well as the top view onto the system (c), with indenter atoms not depicted. In all cases, there is a slow decay of von Mises stresses with the increasing distance from the indenter tip. Both the strained Si (100) surface as well as bottom layer atoms, with a restricted motion during the simulation can be observed in this presentation form.

servations were made by MD studies of Gannepalli et al. [14] using the Stillinger-Weber potential and Lin et al. [154] who also used the Tersoff potential as applied in this study.

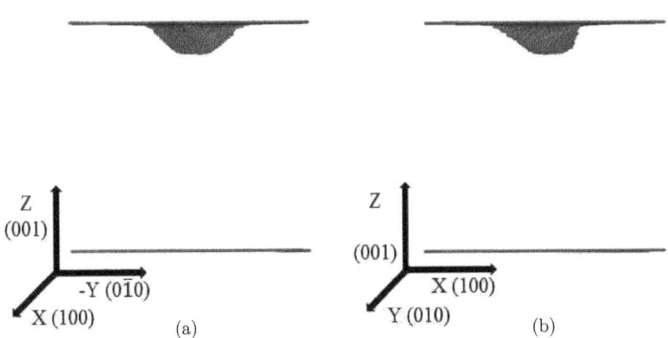

Figure 10.9: Amorphisation of the Si (100) single crystal beneath the indenter tip. In both cases, deviation from the average potential energy was used for the structure identification. In (a) the perspective is set parallel to one of the three edges of the Berkovich tip while in (b) the view direction is perpendicular to one of the three edges.

All previous considerations are made under the assumption of an increasing external loading. The exact analysis of the unloading part of the nanoindentation is important for distinguishing between the elastic and plastic behaviour of the material and the evaluation of the form recovery. As presented in the load-penetration depth diagram, **Figure 10.3**, global stress states of the loading and the unloading curve are very close to each other for all crystal orientations, indicating a great amount of form recovery.

For the study of local distributions, the extend of the amorphised region, the distribution of the potential energy as well as the von Mises stress distribution are presented. **Figure 10.10** presents the spacial extent of the nanoindentation-induced amorphisation upon complete unloading. In direct comparisson with **Figure 10.9** the amorphised region seems to be almost completely vanished, only a small region around the

Chapter 10. Nanoindentation of Si, β-SiC and α-Si$_3$N$_4$ single crystals

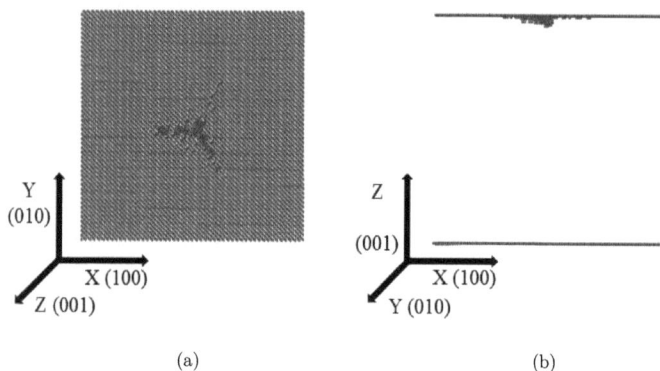

(a)   (b)

Figure 10.10: Nanoindentation-induced amorphisation of the Si (100) single crystal upon unloading. In (a) the top view onto the Si (100) surface is presented, revealing an narrow lateral extension of the amorphous region. From (b) it can be observed that the depth of the amorphised region upon unloading is also very shallow, indicating a high amount of elastic response of silicon towards nanoindentation. Further indication for an elastic response can be viewed trough the analysis of potential energies, **Figure 10.11** and von Mises stresses, **Figure 10.12**.

retracted indenter tip still shows an unordered structure. **Figure 10.11** depicts the potential energy distribution within the silicon sample upon unloading, corresponding to **Figure 10.4** at the maximum indentation depth.

As shown in **Figure 10.11**, only a minor region directly under the indeter tip remains affected upon unloading and has not fully recovered. The same obseravtion is made trough the analysis of von Mises stresses upon unloading presented in **Figure 10.12**. It is allowed to assume that this region would completely vanish under thermal annealing treatement, or will not form at all if the nanoindentation is performed at higher temperatures or at smaller indenter propagation velocities. For the representation of the thermal annealing process of silicon, the reader is referred to **Section 9.2.2**.

Section 10.2. Nanoindentation of silicon

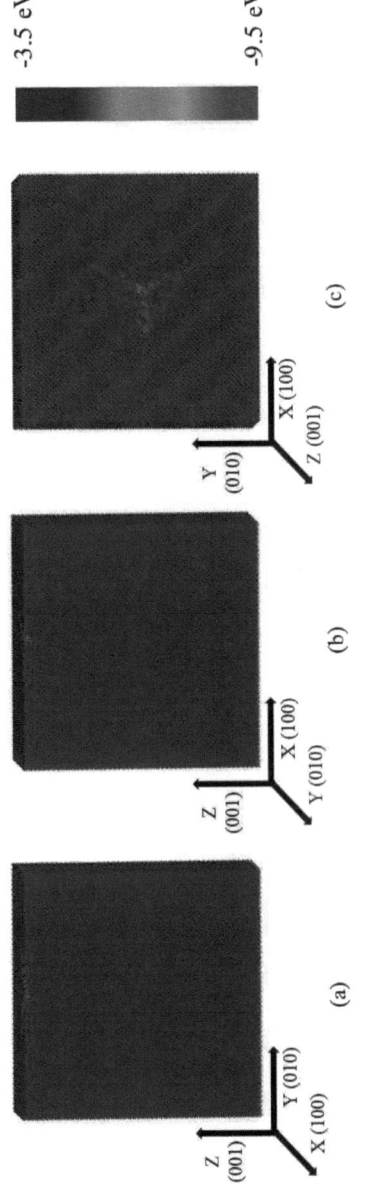

Figure 10.11: Potential energy distribution within the Si (100) substrate upon unloading. In direct comparison with **Figure 10.4**, only a small part of the silicon structure which was directly beneath the indenter tip remains disturbed, here indicated in yellow color, while the major part of atoms obtains their equilibrium positions with a low potential energy (red color).

Chapter 10. Nanoindentation of Si, β-SiC and α-Si$_3$N$_4$ single crystals

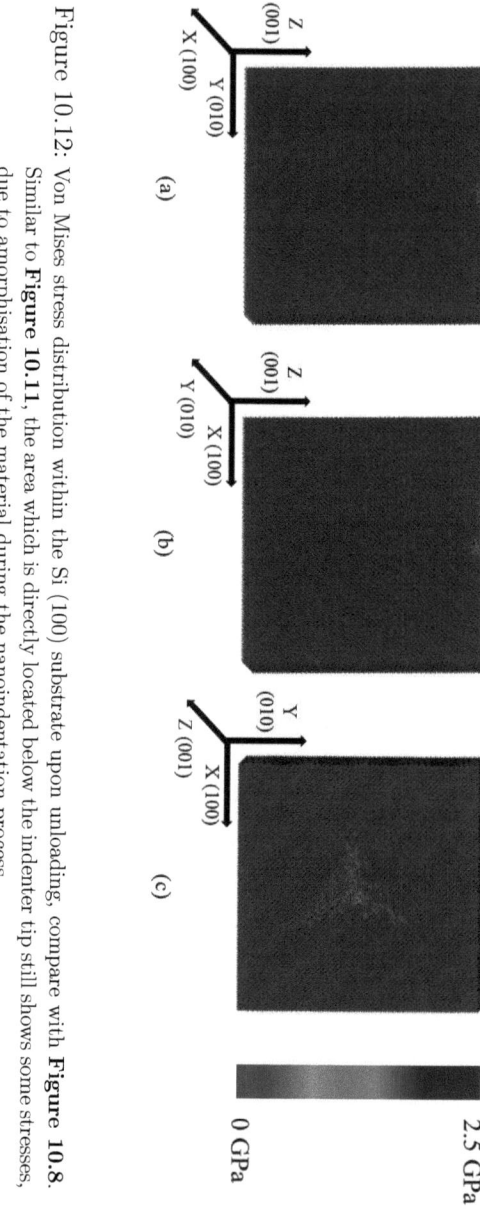

Figure 10.12: Von Mises stress distribution within the Si (100) substrate upon unloading, compare with **Figure 10.8**. Similar to **Figure 10.11**, the area which is directly located below the indenter tip still shows some stresses, due to amorphisation of the material during the nanoindentation process.

10.3 Nanoindentation of β-SiC

In analogy to the nanoindentation of silicon, molecular dynamics simulations of the nanoindentation process of β-SiC were performed. The system under investigation had the same dimensions as in the case of silicon, due to different lattice constants, however, $a(\beta\text{-SiC})=4.32$ Å, the number of particles was different, 2 000 000 in the case of β-SiC. Again, three low-index crystal orientations were chosen, namely β-SiC {100}, {110} and {111}. Berkovich indenter tip of the same size as in the silicon nanoindentation was used, overall the same approach was adopted.

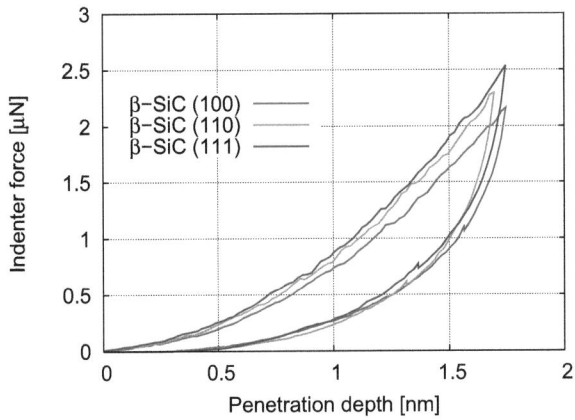

Figure 10.13: Indenter load-indentation depth relation for the nanoindentation of the β-SiC (100), (110) and (111) crystal surface.

Figure 10.13 shows the relation between the indenter load and the indentation depth. As in case of silicon, different crystal orientations show slightly different load-displacement relations. Both in the case of silicon, as also in the case of the structure related β-SiC, the {111} indentation direction shows the highest resistance. These results are in good relation to those obtained by Chen et al. in [61], where also the

Chapter 10. Nanoindentation of Si, β-SiC and α-Si$_3$N$_4$ single crystals

{111} propagation direction in β-SiC was identified as the strongest, although different indenter tip (Vickers) and different interatomic potential (Stillinger-Weber) was used. In both silicon and β-SiC, the {111} direction is parallel to one of four atomic bonds of the diamond tetrahedra. Since greater force is needed to reduce bonds length than for bond bending, observed highest material hardness in case of β-SiC {111} indentation direction is plausible. Overall, the scale of the force axis is around four times higher than in the case of silicon. The corresponding material hardnesses and elastic moduli in dependence on crystal orientation are presented in **Table 10.2**.

Orientation	β-SiC {100}	β-SiC {110}	β-SiC {111}
Hardness [GPa]	43.5	49.3	54.6
E-Modulus [GPa]	2162	2557	2707

Table 10.2: Hardnesses and elastic moduli β-SiC as calculated by molecular dynamics simulation and in dependence on the crystal orientation.

Presented hardnesses are lower than those reported by Szlufarska et al. in [66] where 77 GPa was reported as the hardness of the β-SiC (110) plane. Experimental nanoindentations of 6H-SiC performed by Ziebert and presented in **Chapter 11, Figure 11.8** yields a material hardness of 37 GPa. Taking into account that reported hardness of stoichiometric a-SiC films is 30 GPa [65] and that 27.5 GPa is the hardness of SiC (111), as reported in [61], the calculated hardnesses of single crystal β-SiC fit well into reported literature results. However, calculated elastic moduli are approximately four times higher than those reported elsewhere (2200 GPa vs. 500 GPa).

All three curves in **Figure 10.13** show larger difference between the loading and unloading direction than in case of silicon, **Figure 10.3**, indicating a smaller amount of β-SiC system recovery and higher brittleness. A similar effect was observed for MD simulations of the coating process: While, using thermal annealing, it was possible to cristallize silicon coatings, amorphous SiC coatings retained their structure even after longer annealing times and higher annealing temperatures. This

Section 10.3. Nanoindentation of β-SiC

effect is directly related to the more complex structure of β-SiC, where two atoms form the crystal unit cell.

Next to indenter load-penetration depth relation, local variables such as potential energy and stress distribution were evaluated and are presented in **Figures 10.14-10.18**. The presentation form is the same as in the case of silicon, therefore a comparison between the two related structures appears logical.

Figure 10.14 shows potential energy distribution within the β-SiC (100) system in the case of the greatest calculated indentation depth of 1.7 nm. The presentation form is the same as in case of silicon, in **Figure 10.4**, however, the disturbance is more localized, decaying very fast with increasing distance from the indenter tip.

The analysis of von Mises stresses depicts in an appropriate way the overall stress state of the system and is presented in **Figure 10.15**. A direct comparison with the nanoindentation of silicon reveals the higher localization of von Mises stresses, they decay faster, just few unit cells away from the indenter tip. Again, high tension on the β-SiC crystal surface, as well as the high stress state of imobile atoms representing the substrate holder is visible.

As in the case of silicon, a certain amount of the material amorphisation beneath the indenter tip will form as a result of the nanoindentation process. **Figure 10.16** presents the distribution of the amorphised material at the maximum indentation depth of 1.7 nm from two perspectives perpendicular to the indenter propagation direction. Visible is that the amorphised region is located directly in the vicinity of the indenter tip, its spacial extent is the same as in case of silicon, **Figure 10.9**.

Analysis of the indenter load-penetration depth relation, presented in **Figure 10.13** indicates a smaller amount of system recovery during the unloading part of the nanoindentation process. It is, therefore, interesting to investigate in how far this effect can be observed on the local atomic scale. **Figures 10.17-10.19** enables one to observe this effect. **Figure 10.19** presents the distribution of the amorphised material within the β-SiC (100) substrate upon complete retraction of the inden-

Chapter 10. Nanoindentation of Si, β-SiC and α-Si$_3$N$_4$ single crystals

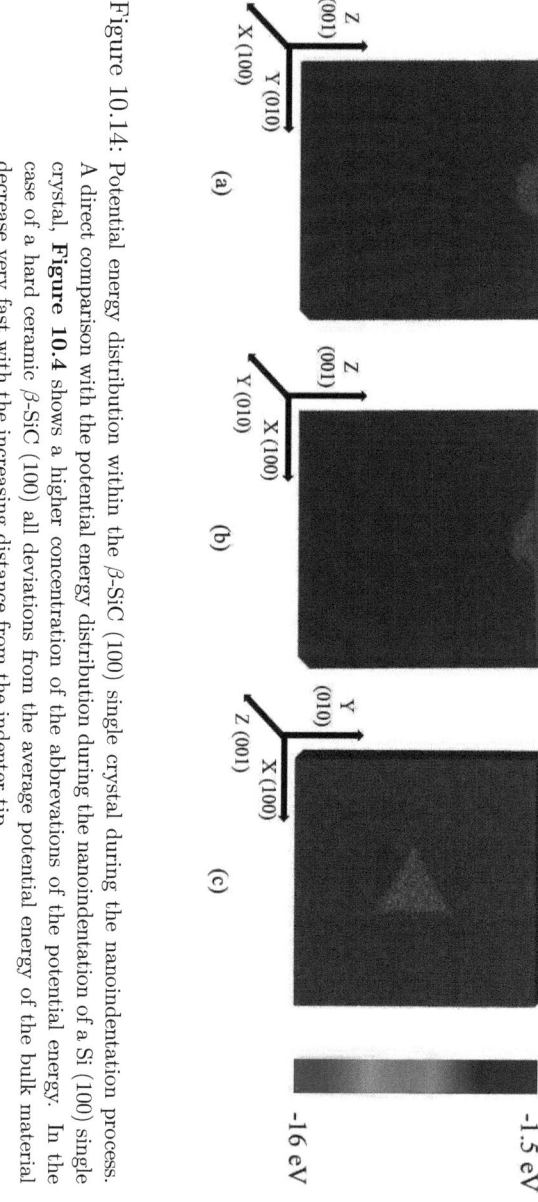

Figure 10.14: Potential energy distribution within the β-SiC (100) single crystal during the nanoindentation process. A direct comparison with the potential energy distribution during the nanoindentation of a Si (100) single crystal, **Figure 10.4** shows a higher concentration of the abbrevations of the potential energy. In the case of a hard ceramic β-SiC (100) all deviations from the average potential energy of the bulk material decrease very fast with the increasing distance from the indenter tip.

Section 10.3. Nanoindentation of β-SiC

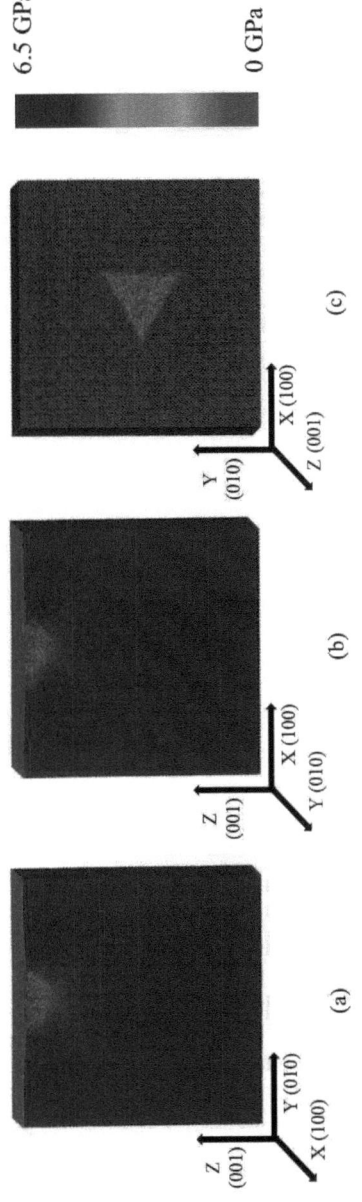

Figure 10.15: Von Mises distribution within the β-SiC (100) substrate during the nanoindentation process. Noticable is the high stress state of strained crystal surface as well as the stress state of imobile atoms representing the substrate holder in an experimental setup. In addition, a comparison with nanoindentation of silicon, **Figure 10.12**, shows a larger localization of von Mises stresses in the vicinity of the indenter tip.

Chapter 10. Nanoindentation of Si, β-SiC and α-Si$_3$N$_4$ single crystals

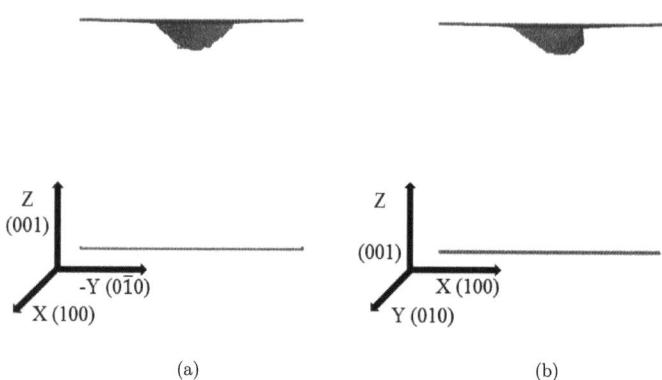

Figure 10.16: Amorphisation of a β-SiC (100) single crystal beneath the indenter tip at the maximum indentation depth. In (a) the perspective is set parallel to one of the three edges of the Berkovich tip while in (b) the view direction is perpendicular to one of the three edges.

ter tip. The amorphised region is significantly larger than in the case of silicon, **Figure 10.10**, althought the maximum indentation depths were the same in both cases. This difference can be directly attributed to the more complex structure of β-SiC. The analysis of potential energies and von Mises stresses further reinforces this statement. **Figure 10.17** presents the potential energy distribution at the end of the unloading phase and $\vec{F}_z=0$. Although the indenter is removed, its imprint is still visible through the potential energies of atoms which were directly below the indenter. Blue colored crystal surface in **Figure 10.17** indicates the presence of unsaturated, two-fold bonded surface atoms and can be ignored. However, dark colored green and blue atoms within the indenter imprint indicate the presence of an amorphous structure just below the indenter tip. This region is also present in the case of silicon nanoindentation but shows there much smaller dimensions, see **Figure 10.11**. Finally, von Mises stress distribution within β-SiC substrate upon unloading, **Figure 10.18**, delivers the best view onto the plasticity of β-SiC due to nanoindentation.

Section 10.3. Nanoindentation of β-SiC

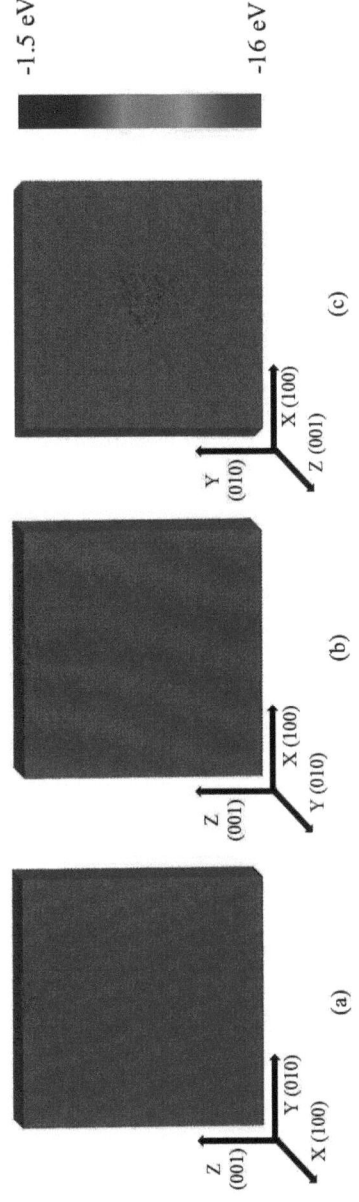

Figure 10.17: Potential energy distribution within the β-SiC (100) substrate upon the unloading phase, compare with the potential energy distribution at the maximum indentation depth in case of β-SiC, **Figure 10.14**, and the nanoindentation of silicon, **Figure 10.11**.

Chapter 10. Nanoindentation of Si, β-SiC and α-Si$_3$N$_4$ single crystals

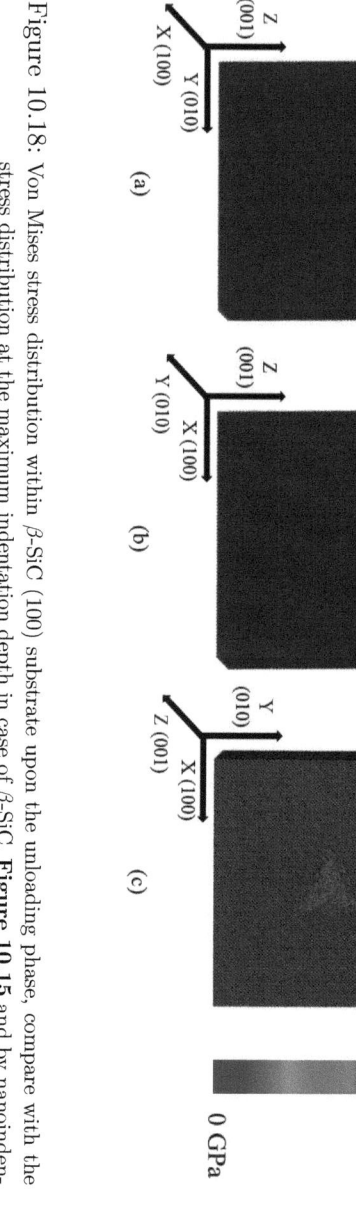

Figure 10.18: Von Mises stress distribution within β-SiC (100) substrate upon the unloading phase, compare with the stress distribution at the maximum indentation depth in case of β-SiC, **Figure 10.15** and by nanoindentation of silicon **Figure 10.12**.

Section 10.4. Nanoindentation of α-Si$_3$N$_4$

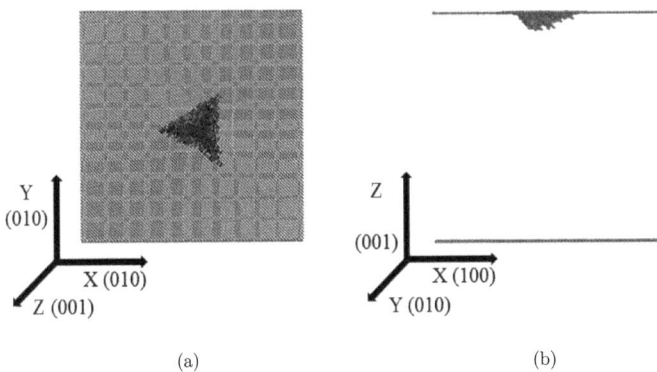

(a)   (b)

Figure 10.19: Nanoindentation-induced amorphisation of a β-SiC (100) single crystal beneath the indenter tip upon unloading. In (a) the top view onto the β-SiC (100) surface is presented, revealing a narrow lateral extension of the amorphous region. It is evident from (b) that the depth of the amorphous region is around 1/3-1/2 of the maximum indentation depth of the Berkovich indenter tip indicating a brittleness of β-SiC. Opposite observations were made for structure-related silicon, **Figure 10.10**.

10.4 Nanoindentation of α-Si$_3$N$_4$

Next to silicon carbide, silicon nitride is also considered as hard ceramic material with desireable mechanical properties. It is therefore the interest to quantify its material properties by a nanoindentation simulation and compare them with both silicon and β-SiC. In addition, previous publications of Walsh et al. [119] on the simulation of nanoindentation of α-Si$_3$N$_4$ (0001) and a-Si$_3$N$_4$ and experimental work of [34], also cited in [119], allows one not only to compare present simulation results with those obtained by modeling or experiment, but also, using a systematic approach, to fill in the missing results of the nanoindentation of other two, mutually perpendicular crystal axes of α-Si$_3$N$_4$, namely $\{10\bar{1}0\}$- and $\{12\bar{1}0\}$-α-Si$_3$N$_4$. For the MD simulation of the nanoindentation

Chapter 10. Nanoindentation of Si, β-SiC and α-Si$_3$N$_4$ single crystals

process of silicon nitride, α-Si$_3$N$_4$ phase was used, having a trigonal structure, presented in **Chapter 7, Section 7.5, Figure 7.24**. On the contrary to the cubic structure of silicon and β-SiC, the trigonal structure of α-Si$_3$N$_4$ is not isotropic, it rather has two different lattice constants defining it, a=5.615 Å and c=7.766 Å. This anisotropy had a great effect onto the sputtering behaviour of α-Si$_3$N$_4$, as presented in **Figure 7.31**. For the simulation of the nanoindentation process, the previousely described Berkovich indenter tip was used, together with a cubic α-Si$_3$N$_4$ system of approx. 30 nm side lenght and 2 000 000 particles. Three low-index crystal orientations of α-Si$_3$N$_4$ were used for the nanoindentation, resulting in the indenter propagation direction parallel to either α-Si$_3$N$_4$ $\{10\bar{1}0\}$, $\{12\bar{1}0\}$ or $\{0001\}$ crystallographic direction. Designated crystallographic directions are perpendicular to each other, the indenter propagation direction was therefore parallel to a-axis (in case of α-Si$_3$N$_4$ $\{10\bar{1}0\}$ orientation), to the triangle height of the same-sided base triangle of the trigonal structure, **Figure 7.24** (for α-Si$_3$N$_4$ $\{12\bar{1}0\}$ orientation) or to the c-axis in the case of α-Si$_3$N$_4$ $\{0001\}$ orientation.

Figure 10.20 presents the resulting indenter load-indentation depth relations. Evident is that for same indentation depths, indenter forces are comparable (slighty lower) to those obtained by the nanondentation of β-SiC, **Figure 10.13**. Compared with silicon, indenter forces in the case of both ceramic materials are approximately four times larger. The resulting material hardnesses and elastic moduli for different crystal orientations of the trigonal α-Si$_3$N$_4$ structure are presented in **Table 10.3**.

Orientation	α-Si$_3$N$_4$ $\{0001\}$	α-Si$_3$N$_4$ $\{10\bar{1}0\}$	α-Si$_3$N$_4$ $\{12\bar{1}0\}$
Hardness [GPa]	29.7	30.0	27.3
E-Modulus [GPa]	915	1008	1060

Table 10.3: Hardnesses and elastic moduli of different crystal orientations of α-Si$_3$N$_4$, calculated by molecular dynamics nanoindentation simulations.

All three curves of the **Figure 10.20** show plasticity (difference be-

Section 10.4. Nanoindentation of α-Si$_3$N$_4$

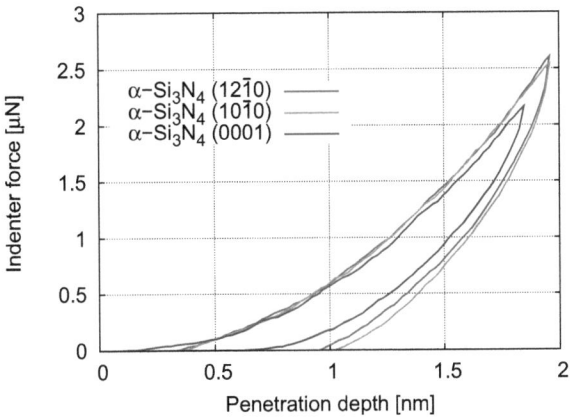

Figure 10.20: Indenter load-displacement relation for 3 low-index crystal orientatons of an α-Si$_3$N$_4$ single crystal.

tween the loading and unloading direction) which is higher than in the case of silicon, but lower than for β-SiC. In addition, α-Si$_3$N$_4$ (0001) oriented system shows the smallest amount of plasticity in between all α-Si$_3$N$_4$ orientations: c-axic, being the longer axis of the trigonal structure, can accumulate a larger amount of elastic energy and release it upon unloading.

In **Figure 10.21** the amorphisation of the Si$_3$N$_4$ material beneath the indenter tip in the case of the maximum indentation depth of 1.8 nm. Visible is the amorphised material in the vicinity of the indenter tip, following closely the shape of the Berkovich tip. Similar observations were made for silicon, **Figure 10.9** and β-SiC, **Figure 10.16**.

Representation of a chemical bond by a Tersoff potential is an approximation which is valid for highly covalent bonds such as present in diamond and silicon single crystals. In the case of SiC an e.g. BN this approximation is still valid due to low ionicity of the bond, see **Section 4.7**. According to [22], the ab initio calculations of the asymmetry

Chapter 10. Nanoindentation of Si, β-SiC and α-Si$_3$N$_4$ single crystals

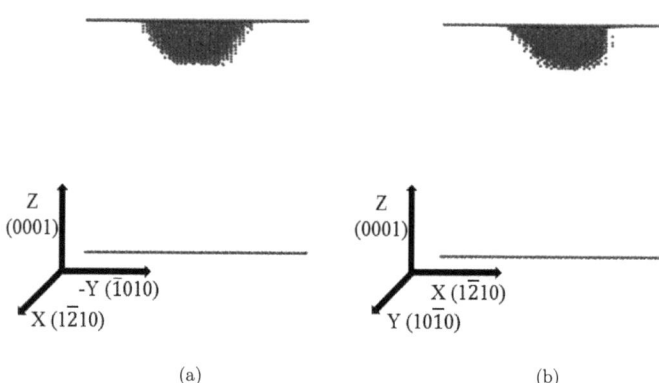

(a) (b)

Figure 10.21: Nanoindentation-induced amorphisation of the α-Si$_3$N$_4$ (0001) single crystal beneath the indenter tip. In both cases, deviation from the average potential energy was used for the structure recognition. In (a) the perspective is parallel to one of the three edges of the Berkovich indenter tip while in (b) the view direction is perpendicular to one of the three edges.

of charge distribution along the Si-N bond, see **Figure 4.7**, can not be neglected. Doing so in the framework of the Tersoff potential leads to results presented in **Figures 10.22** and **10.23** where neither the potential energy nor the distribution of von Mises stresses shows a homogeneous character, such it is the case for silicon and silicon-carbide. Globally, the material hardness of α-Si$_3$N$_4$ corresponds well to those obtained by Walsh et al. in [119], where a Stillinger-Weber based interatomic potential was used.

Section 10.4. Nanoindentation of α-Si$_3$N$_4$

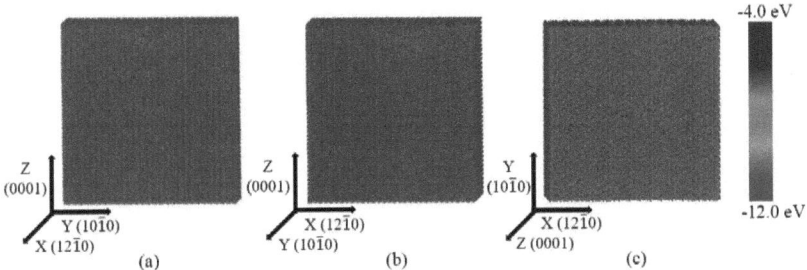

Figure 10.22: Potential energy distribution within the α-Si$_3$N$_4$ (0001) substrate during the nanoindentation process, compare with nanoindentation of silicon, **Figure 10.4** and nanoindentation of β-SiC, **Figure 10.14**.

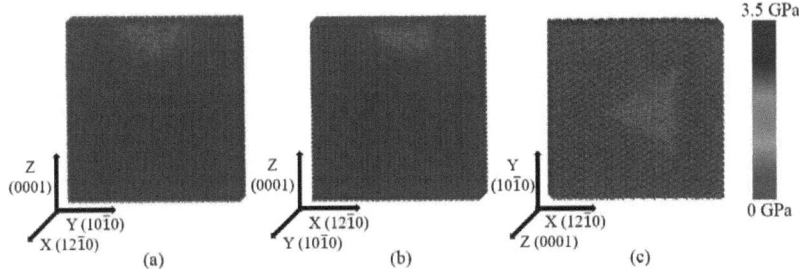

Figure 10.23: Von Mises distribution within the α-Si$_3$N$_4$ (0001) substrate during the nanoindentation process, compare with nanoindentation of silicon, **Figure 10.12**.

Figure 10.25 presents the nanoindentation-induced amorphisation of the Si$_3$N$_4$ material. Although the indenter tip is fully retracted, the material previousely beneath the tip remains amorphous due to brittleness of Si$_3$N$_4$. The vertical extend of the amorphous region is around 1/3-1/2 of the maximum indentation depth, similar to the the nanoindentation of β-SiC, **Figure 10.19**.

Chapter 10. Nanoindentation of Si, β-SiC and α-Si$_3$N$_4$ single crystals

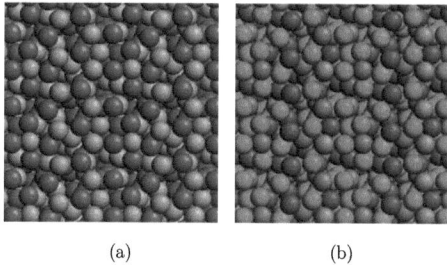

(a)   (b)

Figure 10.24: Comparison of the structure of α-Si$_3$N$_4$ system (silicon atoms in yellow and nitrogen atoms in blue color) and the von Mises stress distribution (low stress in red and high stress in green color). Due to limitations of the Tersoff potential for α-Si$_3$N$_4$, the stress appears to be concentrated on nitrogen atoms rather than be distributed homogenousely over a larger volume.

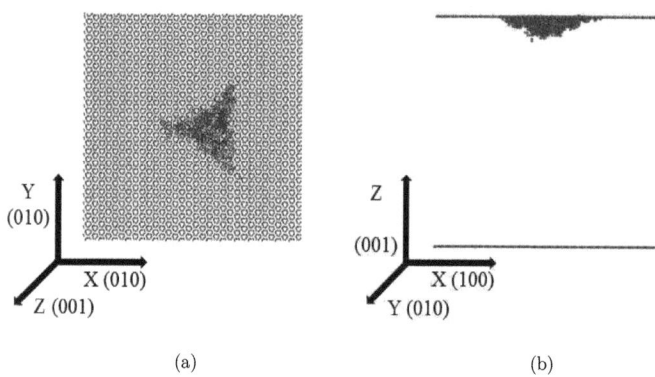

(a)   (b)

Figure 10.25: Nanoindentation-induced amorphisation of the α-Si$_3$N$_4$ (0001) single crystal beneath the indenter tip upon unloading. In (a) the top view onto the α-Si$_3$N$_4$ (0001) surface is presented, revealing an narrow lateral extension of the amorphous region. From (b) it can be observed that the depth of the amorphised region is around 1/3-1/2 of the maximum indentation depth of the Berkovich indenter tip indicating a brittleness of α-Si$_3$N$_4$. Similar observation is made for silicon carbide, **Figure 10.19**.

10.5 Summary

Molecular dynamics simulations of the nanoindentation process of silicon, β-SiC and α-Si$_3$N$_4$ single crystals were performed. Tersoff potential was used for the description of interatomic bonds in the designated material system, while the interaction with atoms of the Berkovich indenter tip was represented trough a short-range pair potential presented earlier in the **Section 4.6.1** of this work.

For Si (100), (110) and (111) crystal surfaces, hardnesses were determined to be 7.8 GPa, 8.3 GPa and 9.4 GPa, which corresponds well to the experimentally obtained hardness of Si (100) of 11 GPa presented in **Figure 11.9**. In addition, elastic moduli of the three silicon surfaces were estimated to 178 GPa, 270 GPa and 224 GPa, also in good agreement with experimental results of 200 GPa for Si (100) surface. The indentation load-displacement relation for all three crystal orientation of silicon indicates a large amount of elastic response of silicon, this could be confirmed by the analysis of the local variables such as potential energy and stress distribution as well as the analysis of the amorphous region remaining upon unloading.

The same approach was adopted on the nanoindentation of β-SiC. Hardnesses of 43.5 GPa, 49.3 GPa and 54.6 GPa for β-SiC (100), (110) and (111) surfaces could be determined. It was observed, that in both silicon and β-SiC, the (111) crystal surface shows the highest hardness, which is in good relation to observations reported in the literature. However, calculated elastic moduli of β-SiC are several times higher than 748 GPa, calculated by MD in [6], or experimentally measured by Carlos Ziebert and presented in **Figure 11.9**. While in case of silicon, the amorphised region disappears completely upon unloading, **Figures 10.9** and **10.10**, a significant part of the material remains amorphous upon a complete unloading of β-SiC, **Figures 10.16** and **10.19**. The vertical extent of the amorphised region is around 1/3-1/2 of the maximum indentation depth of the Berkovich indenter tip.

Mechanical properties of a single crystal α-Si$_3$N$_4$ were also investi-

Chapter 10. Nanoindentation of Si, β-SiC and α-Si$_3$N$_4$ single crystals

gated. Nanoindentation simulations of (0001), (10$\bar{1}$0) and (12$\bar{1}$0) crystal surfaces were performed. Hardnesses of 29.7 GPa, 30.0 GPa and 27.3 GPa could be determined, these are in good agreement with experimental results presented in [34]. Nanoindentation experiments performed by Carlos Ziebert on silicon nitirde coatings yielded hardnesses between 20 GPa and 30 GPa, however, the investigated coatings were amorphous. Both the calculated elastic modulus of 400 GPa, presented in [119] and the elastic moduli of amorphous silicon nitride coatings of approximatelly 250 GPa-350 GPa, presented in **Figure 11.9**, are everestimated by own simulation results, yielding values of around 1000 GPa.

For all three material systems, a large correlation of the distance between the loading and the unloading curve and the size of the amorphous region upon unloading, could be determined, indicating the material ductulity or brittleness. While silicon shows a very ductile nature, **Figures 10.3, 10.9, 10.10**, silicon carbide (**Figures 10.13, 10.16, 10.19**) and silicon nitride (**Figures 10.20, 10.21, 10.25**) are good examples of brittle ceramic materials.

11 Nanoindentation of SiC- and Si$_3$N$_4$-coatings

The overall scope of this work was the modeling and the experimental validation of all stages of the deposition process of Si, SiC and Si$_3$N$_4$ on Si substrates using the method of physical vapour deposition (PVD). On the target side, this included the modeling and the experimental investigation of the sputtering process of Si, β-SiC and α-Si$_3$N$_4$. On the substrate side, the deposition process itself was modeled by molecular dynamics (MD) and the results were compared by experimental results of the PVD deposition process. The analysis of coatings produced in this way included, among other methods, X-ray diffractometry (XRD), presented in **Figures 9.14(a)** and **9.15(a)** as well as atomic force microscopy (AFM), presented in **Figure 9.13**. Finally, coating hardness was evaluated by the method of nanoindentation.

In analogy to the MD modeling of the sputtering and the deposition process, it is also possible to represent the experimental nanoindentation within an MD simulation. In previous sections this was done for Si, β-SiC and α-Si$_3$N$_4$, on the example of differently oriented single crystals. This excursion was done not only to evaluate their material properties, but also to test the limitations of the numerical method itself and to compare results with available literature data, both experimental and those from computer simulations. As shown in **Chapter 10**, it is possible to represent both the macroscopic material properties, such as material hardness and the elastic modulus, as well as different effects which occur on the local atomic scale and which are no longer accessible by macroscopic experimental methods.

Chapter 11. Nanoindentation of SiC- and Si_3N_4-coatings

As a conclusion, it is necessary to evaluate the mechanical properties of SiC- and Si_3N_4-coatings modeled by MD simulations and produced in experiments. This interesting part of the project was outsourced into a student research project. Both MD simulations and the experimental part of the work were done by Ms. Janine Lichtenberg under close supervision of Dr. Carlos Ziebert from the IAM-AWP at the KIT and myself. A detailed description of the approach and the obtained results can be found in [71], for the sake of completeness, a short overview will be given here.

As a model system, stoichiometric SiC- and Si_3N_4-coatings, deposited at 1000 K silicon substrate temperature and presented in **Chapter 9**, **Figures 9.11(d)-9.11(f)** and **9.18(d)-9.18(f)** were investigated. The dependency of the coating structure on the silicon substrate temperature was investigated in the same chapter, both the SiC- and Si_3N_4-coatings show a crystal structure which resembles that of the silicon substrate, the radial distribution functions of the substrate-coating systems are presented in **Figures 9.12** and **9.19**.

In order to represent the nanoindentation process at 0 K, as previousely described for bulk materials, the deposition process was stopped at some stage, the system was stabilized at 1000 K for 10 ns and was cooled down to 0 K during additional 10 ns. This was done by the temperature and pressure control implemented in the *npt axial* numerical integrator, presented in **Section 4.4.2**. This approach resembles the experimental situation by far, the substrate temperature during the deposition process is in the same range as in the simulation presented here and the nanoindentation experiment is performed at a later stage and at room temperature. The development of the averaged potential energy of the SiC-coating/Si-substrate system during the thermal annealing is presented in **Figure 11.1**, while the corresponding radial distribution function at the end of the cooling process is presented in **Figure 11.2(a)** for the SiC-coating/Si-substrate and in **Figure 11.2(b)** for the Si_3N_4-coating/Si-substrate system. It can be observed that the initially ordered structure of the SiC-coating becomes amorphous during the annealling process. The reason for this is the low thickness of the deposited coating. Due to a lattice mismatch of the Si/β-SiC of approx. 20 %, the SiC-coating structure is strained at the coating-substrate interface, this state is unstable from the energetic point of view. In order to represent the process of nanostabilization and a gradual transfer towards the SiC lattice constant within the strained, stoichiometric SiC-coating, larger coating thicknesses, such as those obtained in experiment, see **Figures 9.13, 9.14** and **9.15** in **Chapter 9**, are necessary.

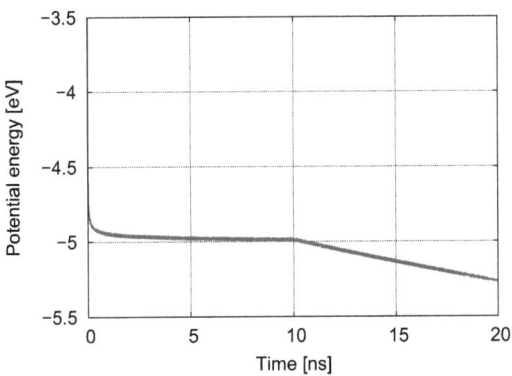

Figure 11.1: Development of the averaged potential energy during the annealing process of a Si-substrate/SiC-coating system. An instant decrease of the average potential energy in the beginning indicates an amorphisation of the SiC-coating structure from the one presented in **Figures 9.11(d)-9.11(f)** and **9.12** to the one presented in **Figure 11.3** with the radial distribution function as in **Figure 11.2(a)**.

Figure 11.3 shows the Si-substrate/SiC-coating system at the end of the annealing process. The substrate-coating interface can be observed in the blue rectangle. In Ziebert et al. [28], the nanoindentation process of a SiC-coating on a silicon substrate was described.

Chapter 11. Nanoindentation of SiC- and Si_3N_4-coatings

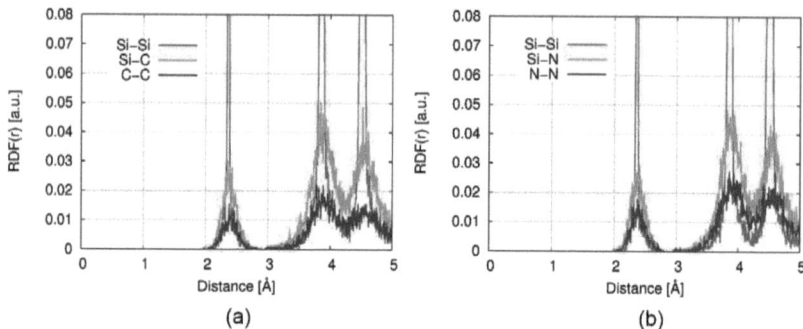

Figure 11.2: Radial distribution function (RDF) of an amorphous Si-substrate/SiC-coating (a) and Si-substrate/Si_3N_4-coating (b) system. Broadening of the RDF-s can be observed in a direct comparison with **Figures 9.12** and **9.19, Chapter 9**. The presented systems were investigated by a nanoindentation simulation, the measured hardnesses are presented in **Figure 11.4**.

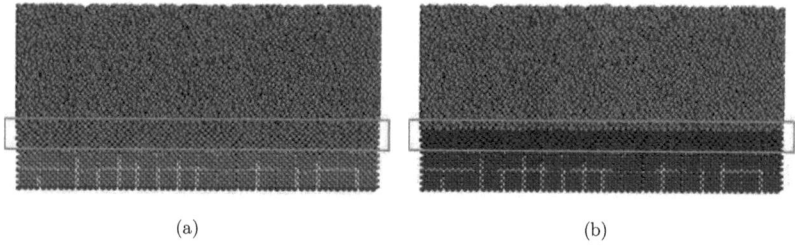

Figure 11.3: SiC-coatings deposited on a Si (100) substrate as in **Figures 9.11(d)-9.11(f)**, annealed at 1000 K for 10 ns and cooled down to 0 K for additional 10 ns, see **Figure 11.1**. Grey dots represent carbon atoms of the SiC-coating and yellow dots represent silicon atoms. Silicon atoms with restricted motion are presented in purple color. On the left, only the motion of bottom layer atoms is restricted in order to prevent the centre of mass drift of the system during the nanoindentation process. On the right, the motion of all substrate atoms is restricted, representing a rigid substrate of an infinite hardness. The substrate-coating interface is presented the blue frame.

During the nanoindentation, the maximum penetration depth of the indenter tip did not exceed 160 nm, which is less that 10 % of the total SiC-coating thickness of 2 μm. This was done in order to minimize the influence of the substrate onto the measurement. Within the presented MD model, this approach can not be used due to the low coating thickness of 10 nm and the substrate-coating ratio of only 2:1 (175:1 in experiment), therefore the substrate influence can not be neglected. In order to estimate the effects of the silicon substrate onto the hardness measurement, between two distinct cases, presented in **Figures 11.3(a)** and **11.3(b)** has to be distinguished. In the first case, **Figure 11.3(a)** a normal behaviour of the silicon substrate atoms was allowed while in the second one, the movement of the silicon substrate atoms was completely restricted, representing a rigid body of infinite hardness.

The measured material hardness is therefore in the first case a property of the total coating-substrate system, while in the second case only the coating hardness is measured. In addition, the hardness of the coating measured in the second case will be an overestimation of the actual value due to low coating thickness.

Figure 11.4 is a graphical representation of coating hardnesses measured by a molecular dynamics simulation for the SiC- and Si_3N_4-coating, with the distinguished behavior of the silicon substrate.

Figure 11.5 presents the hardness of a SiC-coating deposited at different substrate temperatures, from 100 °C to 900 °C and measured with different maximum indenter loads, between 5 mN and 15 mN. Up to 600 °C the deposited SiC-coatings are amorphous, a first nanocrystalline SiC-coating is deposited at 700 °C. This results in lower hardnesses of amorphous coatings, which is measured with all maximum indenter loads. Starting from 700 °C up to 900 °C only nanocrystalline SiC coatings are deposited, however, residual stresses decrease with the increasing substrate temperature. The decrease of residual stresses results in a decreased hardness of high temperature SiC-coatings.

Figure 11.6 presents the experimentally measured hardness of SiC-coatings deposited at 100 °C and 900 °C silicon substrate temperature

Chapter 11. Nanoindentation of SiC- and Si$_3$N$_4$-coatings

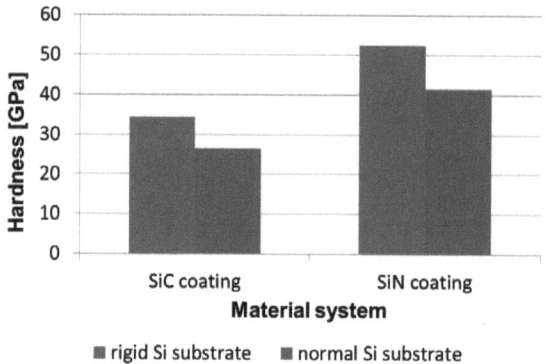

Figure 11.4: Hardness of SiC- and Si$_3$N$_4$-coatings obtained by a nanoindentation simulation. Increased hardnesses for both material systems (blue bars) originate from a rigid silicon substrate, while in the case of lower hardness values (red bars) the motion of substrate atoms was not restricted. In this case an overall hardness of the substrate-coating system was measured, in appreciation of J. Lichtenberg [71].

and of Si$_3$N$_4$-coating deposited at 900 °C substrate temperature. Both the low temperature SiC-coating and the Si$_3$N$_4$-coating are amorphous, while the high temperature deposited SiC-coating has a nanocrystalline structure. In all cases, measured hardness increases with the maximum indenter load. This is due to an increasing indentation depth of the indenter and the resulting absence of near-surface defects, see also **Figure 11.7**. The increased hardness of a nanocrystalline SiC-coating at 900 °C can be observed only for larger indenter loads (> 15 mN), this is due to increased surface roughness, see **Figure 9.13** in **Chapter 9**.

Figure 11.7 presents the coating hardness in dependence of the maximum indentation depth for both SiC- and Si$_3$N$_4$-coatings measured by experimental nanoindentation. For both material systems, larger indentation depths results into increased coating hardness. This effect can be explained considering coating surface defects which have a negative influence onto the coating hardness for smaller indentation depths.

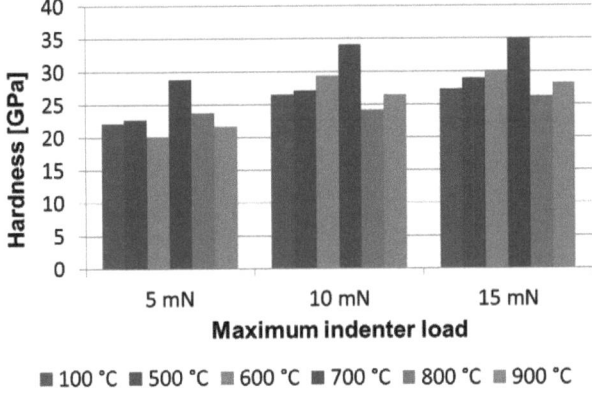

Figure 11.5: Comparison of experimentally measured hardnesses of SiC-coatings (presented in **Figure 9.13**) deposited at different substrate temperatures and measured by different maximum indenter loads. A continuous increase in the coating hardness is evident for substrate temperatures up to 700 °C. This is due to the transition for an amorphous to a nanocrystaline SiC-coating structure, as presented in **Chapter 9**, **Figure 9.13**, in appreciation of J. Lichtenberg [71].

Figure 11.8 presents experimental indenter load-indentation depth relations for different material systems. The experimental nanoindentation of a Si (100) wafer corresponds to the present simulation results, **Chapter 10**, **Figure 10.3**. In both the MD and experimental study, a small amout of plasticity of a Si (100) single crystal is observed, indicating a high amout of form recovery. The experimental nanoindentation of 6H-SiC in **Figure 11.8** is different from the one presented in **Chapter 10**, **Figure 10.13** since the nanoindentation is performed on an α-SiC (instead of β-SiC) as in **Chapter 3**, **Figures 3.4(a)** and **3.4(b)**. In both experimental and MD study of the nanoindentation, a high amout of crystal plasticity is observed for both phases of SiC, which can be deduced from the larger distance between the loading and the unloading curve during the nanoindentation. Compared to the nanoindentation of

Chapter 11. Nanoindentation of SiC- and Si$_3$N$_4$-coatings

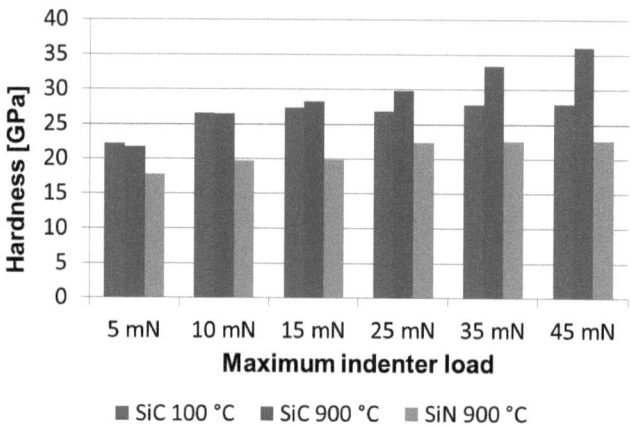

Figure 11.6: Experimentally measured hardnesses of SiC-coatings deposited at 100 °C and 900 °C silicon substrate temperature and of Si$_3$N$_4$-coating deposited at 900 °C substrate temperature, in appreciation of J. Lichtenberg [71].

silicon, the nanoindentation of 6H-SiC shows a force axis which is twice as large for the same indentation depth, in the case of β-SiC, **Chapter 10, Figure 10.13** this ratio is four, corresponding to the increased density of the β-phase of SiC. For all cases of deposited coatings, the resulting hardnesses are between those calculated for Si- and SiC-single crystals. Both amorphous SiC- and Si$_3$N$_4$-coatings have an increased hardness compared with silicon, but a smaller hardness compared with SiC- and Si$_3$N$_4$-single crystals. This is also observed in MD simulations, **Figures 11.4** and **Tables 10.1, 10.2** and **10.3** in **Chapter 10**. The decreased hardness of nanocrystalline β-SiC-coatings in comparison to β-SiC single crystal originates from the gliding of individual grains. The experimental analysis of material properties of an Si- and SiC-wafer, as well as of SiC- and Si$_3$N$_4$-coatings deposited on a silicon substrate under different conditions, are summarized in **Figure 11.9**.

Section 11.1. Summary

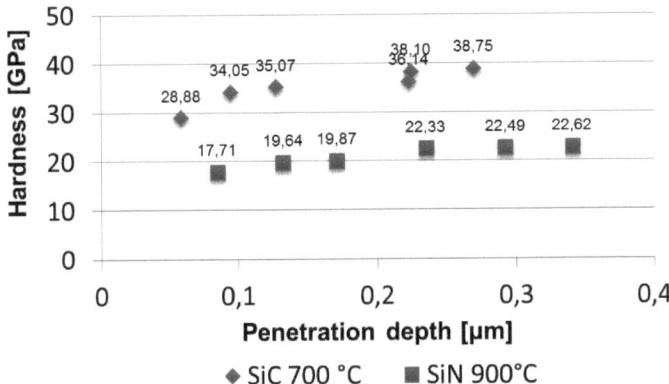

Figure 11.7: Hardnesses of SiC- and Si$_3$N$_4$-coatings measured by experimental nanoindentation and in dependence of the maximum indentation depth. An increased hardness can be observed for larger indentation depths for both material systems due to the absence of near-surface defects, in appreciation of J. Lichtenberg [71].

11.1 Summary

In the presented chapter, molecular dynamics (MD) simulations and experimental investigations of the nanoindentation of deposited SiC- and Si$_3$N$_4$-coating on silicon substrates were described. Molecular dynamics simulations of the nanoindentation were performed examplatory on thin stoichiometric SiC- and Si$_3$N$_4$-coatings, deposited at 1000 K substrate temperature. The behavior of the underlying silicon substrate was either normal or completely rigid. This was done in order to investigate the influence of the relatively weak silicon substrate onto the coating hardness. Experimental nanoindentations were performed by Dr. Carlos Ziebert at the KIT on SiC- and Si$_3$N$_4$-coatings, however, these were deposited at different substrate temperatures, resulting in different structures of the coating material. In addition, the influence of the maximum indentation load as well as maximum indentation depth

Chapter 11. Nanoindentation of SiC- and Si$_3$N$_4$-coatings

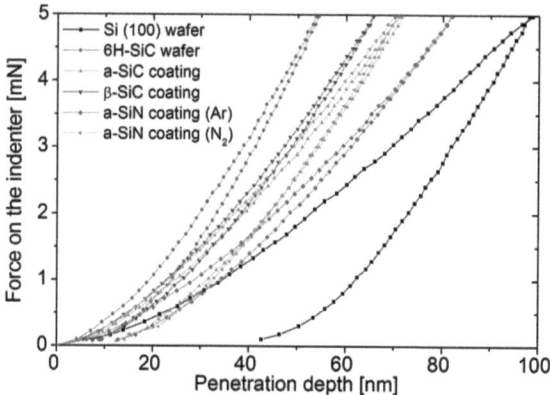

Figure 11.8: Experimental nanoindentation analysis of Si, SiC and Si$_3$N$_4$ as single crystals and coatings, in apprecition of C. Ziebert [27].

was investigated. It can be concluded that, in case of SiC coatings, the measured hardness increases from 18 GPa to 35 GPa in the temperature range from 100 °C to 700 °C and then further decreases to the value of 27 GPa for the substrate deposition temperature of 900 °C. The initial increase is due to the transition from an amorphous to a nanocrystalline coating structure at 700 °C. Above 700 °C, the crystal structure remains nanocrystalline, however, residual stresses decrease, resulting in a decreased coating hardness.

Next to the substrate temperature, the meassured coating hardness is influenced by the nanoindentation parameters themselves. In general, higher maximum indentation depths, related to higher maximum indenter loads, lead to the measurement of increased hardnesses. This effect can be observed in **Figures 11.6** and **11.7**. The differences originate from near-surface defects, reducing the coating hardness in the upper layers. In MD simulations, the hardness of deposited SiC-coatings was 26 GPa for the normal substrate behavior and 34 GPa for the rigid silicon substrate, therefore, the results calculated by MD simulations are

Section 11.1. Summary

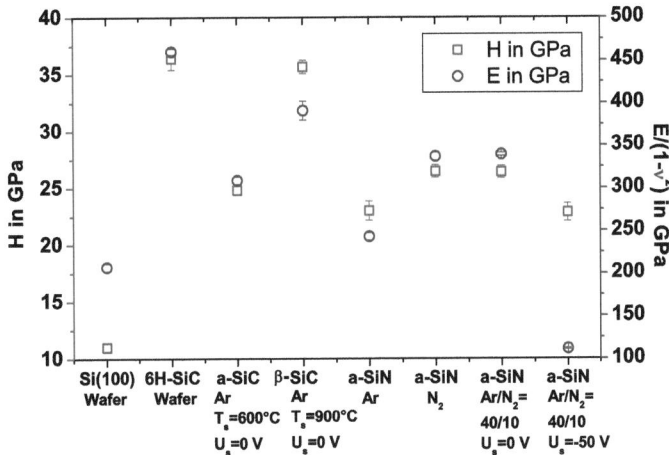

Figure 11.9: Experimental analysis of material hardnesses and elasic moduli of Si, SiC and Si_3N_4 as single crystals and coatings deposited at different deposition parameters, in appreciation of C. Ziebert [27].

in a good agreement with experimental observations.

Regarding Si_3N_4-coatings, the deposition process yielded only amorphous structures independent on the substrate temperature. Depending on the maximum indentation depth, coating hardnesses between 17 GPa and 23 GPa were identified, as presented in **Figure 11.7**. MD simulations of the nanoindentation of Si_3N_4-coatings overestimate the coating hardness by a factor of two, mostely due to the small system size, yielding 41 GPa for the normal substrate behavior and 52 GPa for a rigid silicon substrate. Further optimization of the modeling of the nanoindentation process could be achieved by an increase of the system size, especially, the ratio of the coating and substrate thicknesses (now only 2:1). In addition, an implementation of a better interatomic potential for the Si-N system, representing the partial charge transfer within an Si-N chemical bond is expected to lead to more realistic results.

12 Summary and Conclusions

Silicon- and silicon-based SiC- and Si_3N_4-coatings are currently subject of intensive experimental and empirical research. Especially the mechanical properties of these coatings make them attractive for applications in severe conditions such as high temperatures and/or intensive wear. However, high mechanical and thermal stability of the material system makes the deposition of their thin films a great challenge. Up to now, the deposited SiC and Si_3N_4 thin films showed a predominantly amorphous structure. This is, however, accompanied by inferior mechanical properties than the respective crystalline structure, in addition, the mechanical properties of such films are majorly influenced by the deposition parameters. It is also known, from the related materials system of BN, that nanocrystalline coatings deposited at appropriate process parameters show outstanding thermo-mechanical properties.

The main goal of this work was to combine well established experimental methods of physical vapor deposition (PVD) with modern molecular dynamics (MD) simulations in order to reach the stated objective and deposit nanocrystalline SiC and/or Si_3N_4 thin films on silicon substrates. The experimental PVD process can be roughly divided into two stages, the sputtering process of silicon and carbon or nitrogen atoms by low-energy Ar^+ ions which occurs on the target side and the deposition process itsself, taking place at the substrate side. Finally, the characterisation of the mechanical properties of deposited coatings is an important part of this work. It could be shown that experimental investigations of every stage of the PVD proces can be accompanied by molecular dynamics simulations, hereby providing sinergy effects and a great benefit to the overall investigation.

Chapter 12. Summary and Conclusions

In details, the following milestones could be acomplished:

- Molecular dynamics simulations of the sputtering process of Si, SiC and Si_3N_4 by argon ions. Penetration depths of argon ions in individual target materials could be analyzed as a function of the impact energy and crystal orientation. The overall sputter yield, the sputter yield of individual compounds in the case of SiC and Si_3N_4 as well as the analysis of back sputtered clusters was performed. The simulations data is in an excellent agreement with experimental results of argon etching obtained at by Carlos Ziebert at KIT. Available literature results could be verified, in addition, the same are extended by means of ion impact energy, crystal orientation and analysis regarding stoichiometry of the sputter yield and clustering tendency.

- Deposition proces of Si, SiC and Si_3N_4 coatings on silicon substrates was simulated by molecular dynamics. In the case of compound materials systems, both stoichiometric as well as gradient coatings could be deposited. The major breaktrough in the area of coatings deposition is the deposition of nanocrystalline SiC coatings at 700 °C. No other group has previously reported the deposition of nanocrystalline SiC-coatings at this, relatively low, temperature. Experimental observations made in the deposition process were predicted by molecular dynamics simulations. Simulation results also predict that variing the stoichiometry of SiC and Si_3N_4 coatings during the film growth would lead to a reduction of internal stresses and an increase of the coating-substrate adhesion. However, until now, little experimental attempts have been made in the direction of the deposition of SiC-gradient coatings, making it an interesting topic for a possible continuation of the research in this area.

- Mechanical properties of the deposited SiC- and SiN-coatings were investigated by experimental nanoindentation and MD simulations of the same. A high amount of agreement could be observed for the nanoindentation of SiC-coatings, the calculated hardnesses are in the range of 25-38 GPa both for simulation and experimental results.

Overall a good agreement between molecular dynamics simulations and experimental investigations could be achieved at every stage of the coatings development. This supports the initial idea of combining individual methods in order to obtain synergy effects. Further investigations can therefore be recommended, extending the material system and refining the MD modelling.

13 List of conference contributions and publications

13.1 Conference talks and poster presentations

- A.-P. Prskalo, S. Schmauder, C. Kohler, C. Ziebert, S. Ulrich. J. Ye, *Molecular dynamics simulations of the sputtering of β-SiC by Ar*, poster presentation, Material Science and Engineering (MSE) Congress, September 2008, Nuernberg, Germany.

- A.-P. Prskalo, S. Schmauder, C. Ziebert, J. Ye, S. Ulrich, *Molecular dynamics simulations of the sputtering of SiC and Si_3N_4*, European Material Research Society (EMRS) Spring Meeting, June 2009, Strasbourg, France.

- A.-P. Prskalo, S. Schmauder, C. Ziebert, S. Ulrich. J. Ye, *Molecular dynamics simulations of the sputtering process of Si and the homoepitaxial growth of a Si coating on silicon*, poster presentation, International Workshop of Computational Materials Modelling (IWCMM), September 2009, Constanta, Romania.

- A.-P. Prskalo, S. Schmauder, C. Ziebert, S. Ulrich. J. Ye, *Molecular dynamics simulations of the deposition process of SiC on Si*, oral presentation, Asian-European Plasma Surface Engineering (AEPSE), September 2009, Busan, South Korea.

Chapter 13. List of conference contributions and publications

- A.-P. Prskalo, S. Schmauder, C. Ziebert, S. Ulrich, J. Ye, *Molecular dynamics simulations of the sputtering process of boron nitride*, poster presentation at the International Conference on Metallurgical Coatings & Thin Films (ICMCTF), April 2010, San Diego, United States.

- A.-P. Prskalo, S. Schmauder, C. Ziebert, S. Ulrich, J. Ye, *Molecular dynamics simulations of the sputtering process of $Si/SiC/Si_3N_4$ and the deposition process of SiC on Si*, oral presentation, Collaboratory for Advanced Computing and Simulations (CACS), University of Southern California (USC), Mai 2010, Los Angeles, United States.

- A.-P. Prskalo, S. Schmauder, C. Kohler, *Atomistic simulations of solid solution strengthening of α-iron*, talk at the conference of Mechanics of Composite Materials (MCM) May 2010, Riga, Latvia.

- A.-P. Prskalo, A. M. Siddiq, T. El Sayed, *Molecular dynamics simulations of the dislocation behavior in aluminium*, Computational Solid Mechanics Laboratory (CSML), King Abdullah University of Science and Technology (KAUST), February 2011, Thuwal, Saudi Arabia.

- A.-P. Prskalo, S. Schmauder, C. Ziebert, S. Ulrich, J. Ye, *Molecular dynamics simulations of the sputtering process of Si, β-SiC and α-Si_3N_4 by argon ions*, Reactive sputter deposition (RSD) 2011, December 2011, Linköping, Sweden

- A.-P. Prskalo, J. Lichtenberg, S. Schmauder, C. Ziebert, J. Ye, S. Ulrich, *Molecular dynamics simulation and experimental validation of nanoindentation measurements of silicon carbide coatings*, talk at the International Conference on Metallurgical Coatings & Thin Films (ICMCTF), April 2012, San Diego, United States.

13.2 Journal publications

- C. Ziebert, J. Ye, S. Ulrich, A.-P. Prskalo, S. Schmauder, *Sputter deposition of nanocrystalline β-SiC films and molecular dynamics simulations of the sputter process*, Journal of Nanoscience and Nanotechnology 10, pp. 1-9 (2008).

- A.-P. Prskalo, S. Schmauder, C. Ziebert, J. Ye, S. Ulrich, *Molecular dynamics simulations of the sputtering of SiC and Si_3N_4*, Surface & Coatings Technology 204 (2010) 2081-2084.

- A.-P. Prskalo, S. Schmauder, C. Ziebert, J. Ye, S. Ulrich, *Molecular dynamics simulations of the sputtering process of silicon and the homoepitaxial growth of a Si coating on silicon*, Computational Materials Science 50, pp. 1320-1325 (2011).

- Janine Lichtenberg, *Numerische Modellierung der Nanoindentation an SiC/SiN-Nanolaminaten*, in german, supervised as study research project (Studienarbeit) 2012.

List of Figures

Fig. 3.1	Multiscale simulation of a crack propagation in silicon.	16
Fig. 3.2	Length scales in Multiscale Materials Modeling. .	17
Fig. 3.3	Schematic representation of the Si-C-N material system.	19
Fig. 3.4	Crystal structure of the α- and β-SiC phase. . . .	21
Fig. 3.5	Crystal structure of the α- and β-Si$_3$N$_4$ phase. . .	22
Fig. 4.1	Schematic representation of a two-dimensional periodic system.	32
Fig. 4.2	The indenter-substrate pair potential.	37
Fig. 4.3	Schematic representation of the sp^2 and the sp^3 hybridization states of carbon.	41
Fig. 4.4	Tersoff pair potential part for the sp^2 and the sp^3 hybridization state of carbon.	42
Fig. 4.5	Cut-off function of the Tersoff potential for carbon.	43
Fig. 4.6	Representation of the electronic density for non-polar covalent bonds (H$_2$, Cl$_2$) and a polar covalent bond (HCl).	47
Fig. 4.7	Total valence charge density along the chemical bond in diamond, C$_3$N$_4$, SiC and Si$_3$N$_4$.	50
Fig. 5.1	Processes occuring along the ion trajectory during the ion bombardement	55
Fig. 5.2	Basic mechanisms of back sputtering.	57
Fig. 5.3	Sputter yield as a function of ion impact energy for different atomic masses of the target material.	63

List of Figures

Fig. 5.4	Typical atom ejection patterns in the $(1\bar{1}1)$ plane for Ne$^+$ ion bombardment of the Si (111) surface.	67
Fig. 5.5	Sputter yield of Ge (111) caused by Hg$^+$ ion bombardment.	69
Fig. 5.6	Sputter yield as a function of ion impact energy for the GaAs system.	70
Fig. 6.1	Schematics of the UMIS 2000 ultra microindentation experimental setup	74
Fig. 6.2	Schematics of the most commonly used indenter tips.	75
Fig. 6.3	Nanoindentation of TiC(110)/NbC(110) interface and the resulting σ_{zz} stress distribution.	82
Fig. 6.4	Various models used in nanoindentation simulations of TiC.	83
Fig. 7.1	Argon impact coordinates on a Si(100) crystal surface.	89
Fig. 7.2	Si (100) sputter yield as a function of the argon energy and surface impact coordinate.	90
Fig. 7.3	Forward sputter yield of silicon (100) as function of argon impact energy and impact coordinate.	91
Fig. 7.4	Penetration depth of an argon ion in silicon as a function of impact energy.	93
Fig. 7.5	Sputter yield of silicon as function of argon impact energy calculated with different MD software packages and compared by experimentall results.	94
Fig. 7.6	Silicon cluster sputter yield as function of argon impact energy calculated with IMD.	95
Fig. 7.7	Carbon terminated β-SiC (100) single crystal with designated ion impact coordinates.	97
Fig. 7.8	Back sputter yield of carbon terminated β-SiC(100) crystal surface as function of argon impact energy and surface impact coordinate.	98

List of Figures

Fig. 7.9	Argon penetration depth in β-SiC(100) as function of impact energy and impact coordinate. . .	99
Fig. 7.10	Forward sputter yield of β-SiC(100) as function of argon impact energy and impact coordinate. . . .	100
Fig. 7.11	Top view onto 3 low-index β-SiC crystal surfaces.	101
Fig. 7.12	Differentiated sputter yield of carbon terminated β-SiC (100) as function of argon impact energy. .	103
Fig. 7.13	Differentiated cluster sputter yield of carbon terminated β-SiC (100) as function of argon impact energy.	103
Fig. 7.14	Differentiated sputter yield of silicon terminated β-SiC (100) as function of argon impact energy. .	104
Fig. 7.15	Differentiated cluster sputter yield of silicon terminated β-SiC (100) as function of argon impact energy.	104
Fig. 7.16	Differentiated sputter yield of β-SiC (110) as function of argon impact energy.	105
Fig. 7.17	Differentiated cluster sputter yield of β-SiC (110) as function of argon impact energy.	105
Fig. 7.18	Differentiated sputter yield of carbon terminated β-SiC (111) as function of argon impact energy. .	106
Fig. 7.19	Differentiated cluster sputter yield of carbon terminated β-SiC (111) as function of argon impact energy.	106
Fig. 7.20	Differentiated sputter yield of silicon terminated β-SiC (111) as function of argon impact energy. .	107
Fig. 7.21	Differentiated cluster sputter yield of silicon terminated β-SiC (111) as function of argon impact energy.	107
Fig. 7.22	Comparison of calculated sputter yields of β-SiC for 3 low-index crystal orientations.	108

List of Figures

Fig. 7.23 Comparison of experimental sputter yield values for SiC targets with TRIM [86, 155], TRIDYN and present MD simulations. 109

Fig. 7.24 Schematics of the α-Si$_3$N$_4$ trigonal structure with lattice parameters depicted. 110

Fig. 7.25 Differentiated sputter yield of α-Si$_3$N$_4$(0001) caused by Ar$^+$ impacts, in the low energy range up to 1 keV. 111

Fig. 7.26 Differentiated cluster sputter yield of α-Si$_3$N$_4$(0001) caused by Ar$^+$ impacts, in the low energy range up to 1 keV. 111

Fig. 7.27 Differentiated sputter yield of α-Si$_3$N$_4$(10$\bar{1}$0) caused by Ar$^+$ impacts, in the low energy range up to 1 keV. 112

Fig. 7.28 Differentiated cluster sputter yield of α-Si$_3$N$_4$(10$\bar{1}$0) caused by Ar$^+$ impacts, in the low energy range up to 1 keV. 112

Fig. 7.29 Differentiated sputter yield of α-Si$_3$N$_4$(12$\bar{1}$0) caused by Ar$^+$ impacts, in the low energy range up to 1 keV. 113

Fig. 7.30 Differentiated cluster sputter yield of α-Si$_3$N$_4$(12$\bar{1}$0) caused by Ar$^+$ impacts, in the low energy range up to 1 keV. 113

Fig. 7.31 Comparison of calculated sputter yields of 3 low-index crystal orientations of α-Si$_3$N$_4$. 115

Fig. 8.1 Schematic representation of the magnetron sputtering process 120

Fig. 9.1 Schematic representation of the molecular beam epitaxy (MBE) simulation method 128

Fig. 9.2 Silicon coatings grown on a Si (100) substrate under different deposition conditions. 131

Fig. 9.3	Radial distribution function (RDF) and angular distribution function (ADF) of an amorphous Si-sub-strate/Si-coating system.	132
Fig. 9.4	Radial distribution function (RDF) and angular distribution function (ADF) of a crystalline Si-sub-strate/Si-coating system.	132
Fig. 9.5	Development of the temperature and the averaged potential energy of the Si-substrate/Si-coating system during the thermal annealing process.	134
Fig. 9.6	Development of the structure of the Si-substrate/Si-coating system during a thermal annealing process.	136
Fig. 9.7	Development of the potential energy within the Si-sub-strate/Si-coating system during the annealing process.	137
Fig. 9.8	Development of the temperature within the Si-sub-strate/Si-coating system during the annealing process.	138
Fig. 9.9	Development of von Mises equivalent stresses within the Si-substrate/Si-coating system during the annealing process.	139
Fig. 9.10	Radial (RDF) and angular (ADF) distribution function of the Si-substrate/Si-coating system after the annealing process.	140
Fig. 9.11	Stoichiometric SiC-coatings grown at different substrate temperatures.	143
Fig. 9.12	Radial distribution function (RDF) of a crystalline Si-substrate/ SiC-coating system.	145
Fig. 9.13	Comparison of the surface tomographies made by atomic force microscopy (AFM) of SiC-coatings deposited at different substrate temperatures. . . .	146
Fig. 9.14	Experimental analysis of the temperature dependence of the SiC-coating structure, X-ray diffractogram (XRD) and crystallite size dependence. .	147

List of Figures

Fig. 9.15 Experimental analysis of SiC-coating structure dependence of the bias voltage, X-ray diffractogram and crystallite size dependence. 147

Fig. 9.16 Molecular dynamics simulation of the melting of β-SiC (a) together with the SiC-coating lattice constant, deposited at 900 °C silicon substrate temperature, as a function of bias voltage (b). . . . 148

Fig. 9.17 SiC gradient coatings on a Si (100) substrate: comparison of different coating stoichiometries. 150

Fig. 9.18 Stoichiometric Si_3N_4-coatings grown at different substrate temperatures 152

Fig. 9.19 Radial distribution function of a crystalline Si-sub-strate/-Si_3N_4-coating. 153

Fig. 9.20 SiN gradient coatings on a Si (100) substrate. . . 154

Fig. 9.21 X-ray diffractogram of a an amorphous Si_3N_4-coating on a silicon substrate. 156

Fig. 10.1 Schematic representation of the load-displacement relation. 159

Fig. 10.2 Schematic illustration of the unloading process showing parameters characterizing the contact geometry. 160

Fig. 10.3 Indentation depth vs. indenter load for 3 low-index crystal orientatons of silicon. 162

Fig. 10.4 Potential energy distribution within the Si (100) single crystal during the nanoindentation process. 164

Fig. 10.5 Temperature distribution within the Si (100) single crystal during the nanoindentation. 166

Fig. 10.6 Distribution of normal stresses during nanoindentation of the Si (100) surface. 168

Fig. 10.7 Distribution of shear stresses during nanoindentation of the Si (100) surface. 169

Fig. 10.8 Von Mises stress distribution within the Si (100) substrate during nanoindentation. 170

List of Figures

Fig. 10.9   Amorphisation of the Si (100) single crystal beneath the indenter tip. 171

Fig. 10.10 Nanoindentation-induced amorphisation of the Si (100) single crystal beneath the indenter tip upon unloading. 172

Fig. 10.11 Potential energy distribution within the Si (100) substrate upon unloading 173

Fig. 10.12 Von Mises stress distribution within the Si (100) substrate upon unloading. 174

Fig. 10.13 Indenter load-indentation depth relation for the nano-indentation of the β-SiC (100), (110) and (111) crystal surface. 175

Fig. 10.14 Potential energy distribution within β-SiC (100) substrate during nanoindentation. 178

Fig. 10.15 Von Mises distribution within the β-SiC (100) substrate during nanoindentation. 179

Fig. 10.16 Amorphisation of a β-SiC (100) single crystal beneath the indenter tip at the maximum indentation depth. 180

Fig. 10.17 Potential energy distribution within the β-SiC (100) substrate upon unloading 181

Fig. 10.18 Von Mises stress distribution within β-SiC (100) substrate upon unloading. 182

Fig. 10.19 Nanoindentation-induced amorphisation of a β-SiC (100) single crystal beneath the indenter tip upon unloading. 183

Fig. 10.20 Indenter load-displacement relation for 3 low-index crystal orientatons of an α-Si$_3$N$_4$ single crystal. . 185

Fig. 10.21 Nanoindentation-induced amorphisation of the single crystal in α Si$_3$N$_4$ (0001) crystallographic orientation beneath the indenter tip. 186

List of Figures

Fig. 10.22 Potential energy distribution within the α-Si_3N_4 substrate in (0001) crystallographic orientation during nanoindentation. 187

Fig. 10.23 Von Mises distribution within the α-Si_3N_4 (0001) substrate during nanoindentation. 187

Fig. 10.24 Comparison of the structure of α-Si_3N_4 system and the von Mises stress distribution. 188

Fig. 10.25 Nanoindentation-induced amorphisation of the α-Si_3N_4 (0001) single crystal beneath the indenter tip upon unloading. 188

Fig. 11.1 Development of the averaged potential energy during the annealing process of a Si-substrate/SiC-coat-ing system. 193

Fig. 11.2 Radial distribution function (RDF) of an amorphous Si-substrate/ SiC-coating and Si-substrate/Si_3N_4-coating system. 194

Fig. 11.3 SiC-coatings deposited on a Si (100) substrate, annealed at 1000 K for 10 ns and cooled down to 0 K for additional 10 ns. 194

Fig. 11.4 Hardness of SiC- and Si_3N_4-coatings obtained by a nanoindentation simulation. 196

Fig. 11.5 Comparison of experimentally measured hardnesses of SiC-coatings deposited at different substrate temperatures and measured by different maximum indenter loads. 197

Fig. 11.6 Experimentally measured hardnesses of SiC-coatings deposited at 100 °C and 900 °C silicon substrate temperature and of Si_3N_4-coating deposited at silicon substrate temperature of 900 °C. 198

Fig. 11.7 Hardnesses of SiC- and Si_3N_4-coatings measured by experimental nanoindentation and in dependence of the maximum indentation depth. 199

List of Figures

Fig. 11.8 Experimental nanoindentation analysis of Si, SiC and Si_3N_4 as single crystals and coatings. 200

Fig. 11.9 Experimental analysis of material hardnesses and elastic moduli of Si, SiC and Si_3N_4 as single crystals and coatings deposited at different deposition parameters. 201

List of Tables

Tab. 3.1	Material properties of silicon compared with SiC and Si_3N_4.	19
Tab. 3.2	Comparison of some relevant properties of SiC and other important wide-gap semiconductor materials.	20
Tab. 4.1	Tersoff potential parameters for the calculated quaternary Si-C-N-B system.	45
Tab. 4.2	Phillips (f_i), Pauling (f_i^P) and Harrison ionicities (f_i^H) for number of group-IV, III-V and II-VI-semi-conductors.	49
Tab. 10.1	Silicon hardnesses and elastic moduli as calculated by molecular dynamics simulations and in dependence on the crystal orientation.	163
Tab. 10.2	Hardnesses and elastic moduli β-SiC as calculated by molecular dynamics simulation and in dependence on the crystal orientation.	176
Tab. 10.3	Hardnesses and elastic moduli of different crystal orientations of α-Si_3N_4, calculated by molecular dynamics nanoindentation simulations.	184

Bibliography

[1] Electron microprobe.
http://en.wikipedia.org/wiki/Electron_microprobe.

[2] Finite Element Method (FEM).
http://en.wikipedia.org/wiki/Finite_element_method.

[3] Hexamethyldisilazane (HMDS).
http://en.wikipedia.org/wiki/Bis(trimethylsilyl)amine.

[4] Kohlenstoff.
http://de.wikipedia.org/wiki/Kohlenstoff.

[5] Si - Silicon material properties. IOFFE Physico-Technical Institute,
http://www.ioffe.ru/SVA/NSM/Semicond/Si/mechanic.html.

[6] SiC - Silicon carbide material properties. IOFFE Physico-Technical Institute,
http://www.ioffe.ru/SVA/NSM/Semicond/SiC/mechanic.html.

[7] Silicon carbide.
http://en.wikipedia.org/wiki/Silicon_carbide.

[8] Silicon nitride.
http://en.wikipedia.org/wiki/Silicon_nitride.

[9] **E**xtensible **S**imulation **P**ackage for the **R**esearch on **S**oft Matter.
http://espressomd.org/.

Bibliography

[10] Large-scale Atomic/Molecular Massively Parallel Simulator. http://lammps.sandia.gov/.

[11] CRC Materials Science and Engeneering Handbook. page 471. CRC Boca Raton, 2000.

[12] A. Arsenlis, W. Cai, M. Tang, M. Rhee, T. Oppelstrup, G. Hommes, T.G. Pierce, V.V. Bulatov. Enabling strain hardening simulations with dislocation dynamics. *Modelling and Simulation in Materials Science and Engineering*, 15(6):553–595, 2007.

[13] A. Belger, B. Wolf, T. Sebald, T. Boettger, P. Paufler, H. Mai, E. Beyer. Structural and mechanical characterisation of TiC/VC multilayers using XRD, polarized EXAFS and nanoindentation. *Acta Crystallographica*, A58 (Supplement), C41, 2002.

[14] S. A. Gannepalli. Molecular dynamics studies of plastic deformation during silicon nanoindentation. *Nanotechnology*, 12:250–257, 2001.

[15] A. Nakano, R.K. Kalia, P. Vashishta. Dynamics and morphology of brittle cracks: A molecular-dynamics study of silicon nitride. *Phys. Rev. Lett.*, 75(17):3138–3141, 1995.

[16] A. Noreyan, J. Amar, I. Marinescu. Molecular dynamics simulations of nanoindentation of β-SiC with diamond indenter. *Materials Science and Engineering B*, 117(3):235–240, 2005.

[17] A.-P. Prskalo. Mittelschwere Atome in starken Magnetfeldern. Diplomarbeit, Universität Stuttgart, 2007.

[18] A.-P. Prskalo, J. Lichtenberg, S. Schmauder, C. Ziebert, J. Ye, S. Ulrich. Molecular dynamics simulation and experimental validation of nanoindentation measurements of silicon carbide coatings. *Thin Solid Films*, 2012. submitted.

Bibliography

[19] A.-P. Prskalo, S. Schmauder, C. Ziebert, J. Ye, S. Ulrich. Molecular dynamics simulations of the sputtering of SiC and Si_3N_4. *Surface and Coatings Technology*, 204(12-13):2081–2084, 2010.

[20] A.-P. Prskalo, S. Schmauder, C. Ziebert, J. Ye, S. Ulrich. Molecular dynamics simulations of the sputtering process of silicon and the homoepitaxial growth of a Si coating on silicon. *Computational Materials Science*, 50(4):1320–1325, 2011.

[21] A. Streit, R. Zillig, and P. Klein. Differentiable mesoscopic fields in molecular dynamics simulation: construction, dynamics, and coupling of length scales. In P. Gumbsch, editor, *Multiscale Materials Modelling*, 2006.

[22] A.Y. Liu, M.L. Cohen. Structural properties and electronic structure of low-compressibility materials: β-Si_3N_4 and hypothetical β-C_3N_4. *Physical Review B*, 41(15):10727–10734, 1990.

[23] M. Baskes. Atomistic potentials for the molybdenum-silicon system. *Materials Science and Engineering: A*, 261(1-2):165–168, 1999.

[24] B.J. Stanbery, W.S. Chen, R.A. Mickelson, G.J. Collins, K.A. Emery. Silicon nitride anti-reflection coatings for $CdS/SuInSe_2$ thin film solar cells by electron beam assisted chemical vapor deposition. *Solar Cells*, 14(3):289–291, 1985.

[25] C. Bauer. *Superharte, unterschiedlich gradierte PVD-Kohlenstoffschichten mit und ohne Zusätze von Titan und Silizium*. PhD-Thesis, Universität Karlsruhe, 2003.

[26] C.-W. Han, M.K. Han, S.-H. Paek, C.-D. Kim, I.-J. Chung. Thermal annealing effect on amorphous silicon thin-film transistors fabricated on a flexible stainless steel substrate. *Electrochemical and Solid-State Letters*, 10(5):J65–J67, 2007.

Bibliography

[27] C. Ziebert. Nanoindentation of semiconductor materials and protective coatings. Private communications.

[28] C. Ziebert, J. Ye, S. Ulrich, A.-P. Prskalo, S. Schmauder. Sputter deposition of nanocrystalline β-SiC films and molecular dynamics simulations of the sputter process. *Journal of Nanoscience and Nanotechnology*, 10(2):1120–1128, 2010.

[29] C.B. Cooper, R.G. Hart, J.C. Riley. Low energy sputtering yield of the (111) and ($\overline{111}$) faces of GaAs. *Journal of Applied Physics*, 44(11):5183–5184, 1973.

[30] C.E. Morosanu. The preparation, characterisation and applications of silicon nitride thin films. *Thin Solid Films*, 65(2):171–208, 1980.

[31] C.L. Brooks. Computer simulation of liquids. *Journal of Solution Chemistry*, 18(1):99–99, 1989.

[32] C.M. Marian, M. Gastreich. A systematic theoretical study of molecular Si/N, B/N, and Si/B/N(H) compounds and parameterisation of a force-field for molecules and solids. *Journal of Molecular Structure (Theochem)*, 506(1-3):107–129, 2000.

[33] D. Capriulo. Abhängigkeit der mechanischen Eigenschaften amorpher und kristalliner Hartstoffschichten im System Si-C-N von Temperatur, Schichtdicke und Ionenbeschuss beim Magnetronzerstäuben. Abschlussarbeit, Universität Karlsruhe, 2004.

[34] D. Chakraborty, J. Mukerji. Characterization of silicon nitride single crystals and polycrystalline reaction sintered silicon nitride by microhardness measurements. *Journal of Materials Science*, 15(12):3051–3056, 1980.

[35] D. Engel. *Hartree-Fock-Roothaan-Rechnungen für Vielelektronen-Atome in Neutronenstern-Magnetfeldern*. PhD-Thesis, Universität Stuttgart, 2007.

Bibliography

[36] D.-H. Kim, D.H. Kim, K.S. Lee. Molecular dynamics simulation of energetic ion bombardment onto a-Si_3N_4 surfaces. *Journal of Crystal Growth*, 230(1-2):285–290, 2001.

[37] D.-H. Kim, G.-H. Lee, S.Y. Lee, D.H. Kim. Atomic scale simulation of physical sputtering of silicon oxide and silicon nitride thin films. *Journal of Crystal Growth*, 286(1):71–77, 2006.

[38] D. Humbird, D.B. Graves. Ion-induced damage and annealing of silicon. Molecular dynamics simulations. *Pure and Applied Chemistry*, 74(3):419–422, 2001.

[39] D. Molnar, P. Binkele, S. Hocker, S. Schmauder. Atomistic multiscale simulations on the anisotropic tensile behaviour of copper-alloyed α-iron at different states of thermal ageing. *Philosophical Magazine*, 92(5):586–607, 2012.

[40] D.J. Eaglesham, H.J. Gossmann, M. Cerullo. Limiting thickness h_{epi} for epitaxial growth and room-temperature Si growth on Si(100). *Phys. Rev. Lett.*, 65(10):1227–1230, 1990.

[41] D.K. Kim. Nanoindentation, Lecture 1, Basic Principle. Technical report, Department of Materials Science and Engineering, Korea Advanced Institute of Science and Technology KAIST.

[42] E. Bassous, H.N. Yu, V. Maniscalco. Topology of silicon structures with recessed SiO_2. *Journal of Electrochemical Society*, 123:1729–1737, 1976.

[43] E. Busch, D. Haneman. On the electromechanical effect in Si and Ge. *Tribology Letters*, 2(2):199–206, 1996.

[44] E.T. Gawlinski, J.D. Gunton. Molecular-dynamics simulation of molecular-beam epitaxial growth of the silicon (100) surface. *Phys. Rev. B*, 36(9):4774–4781, 1987.

Bibliography

[45] F. de Brito Mota, J.F. Justo, A. Fazzio. Structural properties of amorphous silicon nitride. *Phys. Rev. B*, 58(13):8323–8328, 1998.

[46] F. Rösch. *Atomistic dynamics of crack propagation in complex metallic alloys*. PhD-Thesis, Universität Stuttgart, 2008.

[47] F. Shimojo, I. Ebbsjö, R.K. Kalia, A. Nakano, J.P. Rino, P. Vashishta. Molecular dynamics simulation of pressure induced structural transformation in silicon carbide. *Physical Review Letters*, 84(15):3338–3341, 2000.

[48] F.H. Stillinger, T.A. Weber. Computer simulation of local order in condensed phases of silicon. *Phys. Rev. B*, 31(8):5262–5271, 1985.

[49] G. Abel. Empirical chemical pseudopotential theory of molecular and metallic bonding. *Phys. Rev. B (Condensed Matter)*, 31(10):6184–6196, 1985.

[50] G. Ecke, R. Kosiba, V. Kharlamov, Y. Trushin, J. Pezoldt. The estimation of sputtering yields for SiC and Si. *Nuclear Instruments and Methods in Physics Research B*, 169(1-2):39–50, 2002.

[51] G.J. Sibona, S. Schreiber, R.H.W. Hoppe, B. Stritzker, A. Revnic. Numerical simulation of the production processes of layered materials. *Materials Science in Semiconductor Processing*, 6(1-3):71–76, 2003.

[52] G.K. Wehner. Controlled sputtering of metals by low-energy Hg ions. *Phys. Rev.*, 102(3):690–704, 1956.

[53] G.S. Anderson. Atom ejection studies for sputtering of semiconductors. *Journal of Applied Physics*, 37(9):3455–3458, 1966.

[54] G.S. Anderson, G.K. Wehner. Atom ejection patterns in single crystal sputtering. *Journal of Applied Physics*, 31(12):2305–2313, 1960.

Bibliography

[55] G.S. Anderson, G.K. Wehner. Temperature dependence of ejection patterns in Ge, Si, InSb and InAs sputtering. *Surface Science*, 2(0):367–375, 1964.

[56] G.S. Anderson, G.K. Wehner, H.J. Olin. Temperature dependence of ejection patterns in Ge sputtering. *Journal of Applied Physics*, 34(12):3492–3495, 1963.

[57] H.-J. Gossmann, L.C. Feldman. Initial stages of silicon molecular-beam epitaxy: Effects of surface reconstruction. *Phys. Rev. B*, 32(1):6–11, 1985.

[58] H.-J. Gossmann, L.C. Feldman, W.M. Gibson. The influence of reconstruction on epitaxial growth: Ge on Si(100)-(2×1) and Si(111)-(7×7). *Surface Science*, 155(2-3):413–431, 1985.

[59] H. Kleykamp. Gibbs energy of formation of SiC: a contribution to the thermodynamic stability of the modification. *Bericht der Bunsen-Gesellschaft - Physical Chemistry Chemical Physics*, 102(9):1231–1234, 1998.

[60] H. Lorentz, I. Eisele, J. Ramm, J. Erdinger, M. Büchler. Characterisation of low temperature SiO_2 and Si_3N_4 films deposited by plasma enhanced evaporation. *Journal of Vacuum and Science Technology B*, 9(2):208–214, 1991.

[61] H.-P. Chen, R.K. Kalia, A. Nakano, P. Vashishta, I. Szlufarska. Multimillion-atom nanoindentation simulation of crystalline silicon carbide: Orientation dependence and anisotropic pileup. *Journal of Applied Physics*, 102(6):063514 (1–9), 2007.

[62] H. Suematsu, J.J. Petrovic, T.E. Mitchell. Deformation and toughness of α-silicon nitride single crystals. *Proceedings of MRS 1992 Fall Meeting*, 287:449–454, 1993.

Bibliography

[63] H.P. Löbl, M. Huppertz. Thermal stability of nonstoichiometric silicon nitride films made by reactive DC magnetron sputter deposition. *Thin Solid Films*, 317(1-2):153–156, 1998.

[64] I. Sugimoto, S. Nakano, H. Kuwano. Tensilely-stressed SiN films reactevely sputtered in Kr-N_2 plasmas for producing free-standing devices. *Thin Solid Films*, 268(1-2):152–160, 1995.

[65] A. P. I. Szlufarska, R.K. Kalia. A molecular dynamics study of nanoindentation of amorphous silicon carbide. *Journal of Applied Physics*, 102(2):0235091–0235099, 2007.

[66] I. Szlufarska, R.K. Kalia, A. Nakano, P. Vashishta. Atomistic mechanisms of amorphization during nanoindentation of SiC: A molecular dynamics study. *Phys. Rev. B*, 71:174113 (1–13), 2005.

[67] I. Zarudi, L.C. Zhang, M.V. Swain. Microstructure evolution in monocrystalline silicon in cyclic microindentations. *Journal of Materials Research*, 18(4):758–761, 2003.

[68] J. Comas, C. B. Cooper. Mass-spectrometric study of sputtering of single crystals of GaAs by low-energy Ar ions. *Journal of Applied Physics*, 38(7):2956–2960, 1967.

[69] J. Comas, C.B. Cooper. Sputtering yields of several semiconducting compounds under argon ion bombardment. *Journal of Applied Physics*, 37(7):2820–2822, 1966.

[70] J. Farren, W.J. Scaife. Sputtering of GaAs single crystals by 8-16 keV argon ions. *Talanta*, 15(11):1217–1226, 1968.

[71] J. Lichtenberg. Experimentelle Untersuchung und numerische Modellierung der Nanoindentation an SiC/SiN-Nanolaminaten. Student-Thesis, Universität Stuttgart, 2012.

Bibliography

[72] J. Pezoldt, B. Stottko, G. Kupris, G. Ecke. Sputtering effects in hexagonal silicon carbide. *Materials Science and Engineering: B*, 29(13):94–98, 1995.

[73] J. Stadler, R. Mikulla, H.-R. Trebin. *International Journal of Modern Physics C*, (8):1131–1140, 1997.
http://www.itap.physik.uni-stuttgart.de/~imd/.

[74] J. Tersoff. New empirical model for the structural properties of silicon. *Phys. Rev. Lett.*, 56(6):632–635, 1986.

[75] J. Tersoff. Empirical interatomic potential for carbon, with applications to amorphous carbon. *Phys. Rev. Lett.*, 61(25):2879–2882, 1988.

[76] J. Tersoff. Empirical interatomic potential for silicon with improved elastic properties. *Phys. Rev. B*, 38(14):9902–9905, 1988.

[77] J. Tersoff. New empirical approach for the structure and energy of covalent systems. *Phys. Rev. B*, 37(12):6991–7000, 1988.

[78] J. Tersoff. Modeling solid-state chemistry: Interatomic potentials for multicomponent systems. *Phys. Rev. B*, 39(8):5566–5568, 1989.

[79] J.B. Kortright, D.L. Windt. Amorphous silicon carbide coatings for extreme ultraviolet optics. *Appl. Opt.*, 27(14):2841–2846, 1988.

[80] J.B. Malherbe. Sputtering of compound semiconductor surfaces. I. Ion-solid interactions and sputtering yields. *Critical Reviews in Solid State and Materials Sciences*, 19(2):55–127, 1994.

[81] J.E. Bradby, J.S. Williams, J. Wong-Leung, M.V. Swain, P. Munroe. Mechanical deformation in silicon by micro-indentation. *Journal of Materials Research*, 16(5):1500–1507, 2001.

[82] J.E. Lennard-Jones. On the determination of molecular fields. II. From the equation of state of a gas. *Proceedings of the Royal Society of London. Series A*, 106(738):463–477, 1924.

Bibliography

[83] J.E. Lennard-Jones. Cohesion. *Proceedings of the Physical Society*, 43(5):461–482, 1931.

[84] J.F. Ziegler, J.P. Biersack, U. Littmark. *The Stopping and Range of Ions in Matter*. Pergamon, New York, 1985.

[85] J.H. Kim, K.W. Chung. Microstructure and properties of silicon nitride thin films deposited by reactive bias magnetron sputtering. *Journal of Applied Physics*, 83(11):5831–5839, 1998.

[86] J.P. Biersack. A Monte Carlo computer program for the transport of energetic ions in amorphous targets. *Nuclear Instruments and Methods*, 174(1-2):257–269, 1980.

[87] J.S. Koehler. Attempt to design a strong solid. *Phys. Rev. B*, 2(2):547–551, 1970.

[88] K. Albe. *Computersimulationen zu Struktur und Wachstum von Bornitrid*. PhD-Thesis, Forschungszentrum Rossendorf/Technische Universität Dresden, 1998.

[89] K. Matsunaga, Y. Iwamoto. Molecular dynamics study of atomic structure and diffusion behavior in amorphous silicon nitride containing boron. *Journal American Ceramic Society*, 84(10):2213–2219, 2001.

[90] K. Wasa, T. Nagai, S. Hayakawa. Structure and mechanical properties of r.f. sputtered SiC films. *Thin Solid Films*, 31(3):235–241, 1976.

[91] L. van Dommelen. Physical Interpretation of the Virial Stress. http://www.eng.fsu.edu/~dommelen/papers/virial/mosaic/index.html.

[92] L.G. Parrat. Surface studies of solids by total reflection of X-rays. *Physical Review B*, 95(2):359–369, 1954.

Bibliography

[93] F. Ltd. Materials Explorer. http://www.cache.fujitsu.com/materialsexplorer/index.shtml.

[94] M. Balooch, M. Moalem, W.-E. Wang, A.V. Hamza. Low-energy Ar ion-induced and chlorine ion etching of silicon. *Journal of Vacuum Science & Technology A: Vacuum, Surfaces and Films*, 14(1):229–233, 1996.

[95] M. Born, Th. von Karman. Über Schwingungen in Raumgittern. *Physikzeitschrifft*, 13:297–309, 1912.

[96] M. Griebel, J. Hamaekers. Molecular dynamics simulations of boron-nitride nanotubes embedded in amorphous Si-B-N. *Computational Materials Science*, 39(3):502–517, 2007.

[97] M. Jansen, J.C. Schön, L. van Wüllen. The route to the structure determination of amorphous solids: A case study of the ceramic $Si_3B_3N_7$. *Angewandte Chemie International Edition*, 45(26):4244–4263, 2006.

[98] M. Klews. *Diskretisierungverfahren zur Untersuchung von Atomen in zeitabhängigen elektrischen Felden und in extrem starken Magnetfeldern*. PhD-Thesis, Eberhard-Karls-Universitaet zu Tübingen, 2003.

[99] M. Lattemann. *Herstellung und Charakterisierung kovalent gebundener Ein- und Viellagenschichten aus dem System B-C-N-Si*. PhD-Thesis, Universität Karlsruhe, 2004.

[100] M. Timonova, B.-J. Lee, B.J. Thijsee. Sputter erosion of Si (001) using a new silicon MEAM potential and different thermostats. *Nuclear Instruments and Methods in Physics Research B*, 255(1):195–201, 2007.

[101] M.E. Barone, D.B. Graves. Chemical and physical sputtering of fluorinated silicon. *Journal of Applied Physics*, 77(3):1263–1274, 1995.

[102] M.I. Baskes. Application of the embedded-atom method to covalent materials: A semiempirical potential for silicon. *Phys. Rev. Lett.*, 59(23):2666–2669, 1987.

[103] M.I. Baskes. Modified embedded-atom potentials for cubic materials and impurities. *Phys. Rev. B*, 46(5):2727–2742, 1992.

[104] M.I. Baskes, J.S. Nelson, A.F. Wright. Semiempirical modified embedded-atom potentials for silicon and germanium. *Phys. Rev. B*, 40(9):6085–6100, 1989.

[105] M.P. Allen, D.J. Tildesley. *Computer Simulation of Liquids*. Oxford Science Publications, Clarendon Press, Oxford, 1987.

[106] M.S. Daw, M.I. Baskes. Embedded-atom method: Derivation and application to impurities, surfaces, and other defects in metals. *Phys. Rev. B*, 29(12):6443–6453, 1984.

[107] M.T. Robinson. Theoretical aspects in monocrystal sputtering. In R. Behrisch, editor, *Sputtering and Particle Bombardment I. Chap. 3*, volume 47, pages 73–144. Springer Verlag Berlin/Heidelberg, 1981.

[108] N. Resta. *Molecular dynamics simulatons of presursor-derived Si-C-N ceramics*. PhD-Thesis, Uni-Stuttgart, 2005.

[109] N. Resta, C. Kohler, H.-R. Trebin. Molecular dynamics simulations of amorphous Si-C-N ceramics: Composition dependence of the atomic structure. *Journal of the American Ceramic Society*, 86(8):1409–1414, 2003.

[110] N.A. Kubota, D.J. Economou, S.J. Plimpton. Molecular dynamics simulations of low-energy (25-200 eV) argon ion interactions with

silicon surfaces: Sputter yields and product formation pathways. *Journal of Applied Physics*, 83(8):4055–4063, 1998.

[111] O. Penrose, P.C. Fife. Thermodynamically consistent models of phase-field type for the kinetic of phase transitions. *Physica D: Nonlinear Phenomena*, 43(1):44–62, 1990.

[112] P. Erhart, K. Albe. Analytical potential for atomistic simulations of silicon, carbon, and silicon carbide. *Phys. Rev. B*, 71(3):035211 (1–14), 2005.

[113] P. Kizler, S. Schmauder. Simulation of the nanoindentation of hard metal carbide layer systems - the case of nanostructured ultra-hard carbide layer systems. *Computational Materials Science*, 39(1):205–213, 2007.

[114] P. Mandricci, A. Chiodoni, G. Cicero, S. Ferrero, F. Giorgis, C. F. Pirri, G. Barucca, P. Musumeci, R. Reitano. Heteroepitaxy of 3C-SiC by electron cyclotron resonance - CVD technique. *Applied Surface Science*, 184(1):43–49, 2001.

[115] P. Pröschel, W. Rosner, G. Wunner, H. Ruder, H. Herold. Hartree-fock calculations for atoms in strong magnetic fields. I. energy levels of two-electron systems. *Journal of Physics B: Atomic and Molecular Physics*, 15(13):1959, 1982.

[116] P. Sigmund. Theory of Sputtering. I. Sputtering Yield of Amorphous and Polycrystalline Targets. *Phys. Rev.*, 184(2):383–416, 1969.

[117] P. Sigmund, M. Szymonski. Temperature-dependent sputtering of metals and insulators. *Applied Physics A: Materials Science & Processing*, 33(3):141–152, 1984.

[118] P. Vashishta, R.K. Kalia, A. Nakano, J.P. Rino. Interaction potential for silicon carbide: A molecular dynamics study of elastic con-

Bibliography

stants and vibrational density of states for crystalline and amorphous silicon carbide. *Journal of Applied Physics*, 101(10):103515 (1-12), 2007.

[119] P. Walsh, A. Omeltchenko, R.K. Kalia, A. Nakano, P. Vashishta, S. Saini. Nanoindentation of silicon nitride: a multi-million atom molecular dynamics study. *Applied Physics Letters*, 82(1):118-120, 2003.

[120] P.C. Zalm. Energy dependence of the sputtering yield of silicon bombarded with neon, argon, krypton, and xenon ions. *Journal of Applied Physics*, 54(5):2660-2666, 1983.

[121] P.C. Zalm. Some useful yield estimates for ion beam sputtering and ion plating at low bombarding energies. *Journal of Vacuum Science & Technology B: Microelectronics and Nanometer Structures*, 2(2):151-152, 1984.

[122] P.H. Morse. Diatomic molecules according to the wave mechanics. II. Vibrational levels. *Phys. Rev.*, 34(1):57-64, 1929.

[123] Q. Wahab, L. Hulmtan, I.P. Ivanov, M. Wilander, J.-E. Sundgren. Growth of and characterisation of 3C-SiC films on Si substrates by reactive magnetron sputtering; effects of CH_4 partial pressure on the crystalline quality, structure and stoichiometry. *Thin Solid Films*, 261(1-2):317-321, 1995.

[124] R. Astala, M. Kaukonen, R.M. Nieminen, T. Heine. Nanoindentation of silicon surfaces: Molecular-dynamics simulations of atomic force microscopy. *Phys. Rev. B*, 61(4):2973-2980, 2000.

[125] R. Kosiba. *Auger electron spectroscopy and low-energy ion bombardment of silicon carbide*. PhD-Thesis, Technical University of Ilmenau, 2005.

[126] R. Rurali. *Theoretical studies of defects in silicon carbide*. PhD-Thesis, Universitat Autònoma de Barcelona, 2003.

Bibliography

[127] R.M. Wallace, Y. Wei. Dry oxidation resistance of ultrathin nitride films: Ordered and amorphous silicon nitride on Si (111). *Journal of Vacuum Science and Technology B*, 17:970–977, 1999.

[128] S. Adachi. *Properties of group IV, III-V and II-VI semiconductors*. Wiley Series in Materials for Electronic & Optoelectronic Applications, 2005.

[129] S. Bücheler. *Diffusions-Quanten-Monte-Carlo-Simulationen für Vielelektronen-Atome in Neutronensternmagnetfeldern*. PhD-Thesis, Universität Stuttgart, 2007.

[130] S. Hocker. *Molekulardynamiksimulation der Diffusion in dekagonalen Quasikristallen mit optimierten Wechselwirkungspotentialen*. PhD-Thesis, Universität Stuttgart, 2007.

[131] S. Melchionna, G. Chicotti, B.L. Holian. Hoover npt dynamics for systems varying in shape and size. *Molecular Physics*, 3(78):533–544, 1993.

[132] S. Munetoh, T. Motooka, K. Moriguchi, A. Shintani. Interatomic potential for Si-O systems using Tersoff parameterization. *Computational Materials Science*, (39):334–339, 2007.

[133] S. Nosé. A molecular dynamics method for simulations in the canonical ensemble. *Molecular Physics*, 2(52):255–268, 1984.

[134] S. Sonntag, J. Roth, F. Gähler, H.-R. Trebin. Femtosecond laser ablation of aluminum. *Applied Surface Science*, 2009(255):9742–9744, 2009.

[135] S. Sonntag, J. Roth, H.-R. Trebin. Molecular dynamics simulations of laser induced surface melting in orthorhombic Al_3Co_4. *Applied Physics A*, 2010.

[136] S.A. Awan, R.D. Gould, S. Gravano. Electrical conduction processes in silicon nitride thin films prepared by r.f. magnetron sput-

tering using nitrogen gas. *Thin Solid Films*, 355-356(1):456–460, 1999.

[137] ScoTech. *Principle of DC magnetron sputtering*.

[138] S.M. Foiles, M.I Baskes, M.S. Daw. Embedded-atom-method functions for the fcc metals Cu, Ag, Au, Ni, Pd, Pt, and their alloys. *Phys. Rev. B*, 33(12):7983–7991, 1986.

[139] S.O. Nielsen, R.E. Bulo, P.B. Moore, B. Ensing. Recent progress in adaptive multiscale molecular dynamics simulations of soft matter. *Physical Chemistry Chemical Physics*, 12(39):12401–12414, 2010.

[140] T. Aoki, S. Chiba, J. Matsuo, I. Yamada, J.P. Biersack. Molecular dynamics and Monte-Carlo simulation of sputtering and mixing by ion irradiation. *Nuclear Instruments and Methods in Physics Research Section B: Beam Interactions with Materials and Atoms*, 180(1-4):312–316, 2001.

[141] T.-K. Kim, G.-B. Kim, Y.-G. Yoon, C.-H. Kim, B.-I. Lee, S.-K. Joo. Scanning rapid thermal annealing process for poly-silicon thin film transistor. *Japanese Journal of Applied Physics*, 39(10):5773–5775, 2000.

[142] T. Motooka, S. Munetoh, R. Kishikawa, T. Kuranaga, T. Ogata, T. Mitani. Molecular-dynamics simulations of recrystallization processes in silicon: Nucleation and crystal growth in the solid-phase and melt. *ECS Transactions*, 3(8):207–213, 2006.

[143] T. Serikawa, A. Okamoto. Properties of magnetron-sputtered silicon nitride films. *Jornal of Electrochemical Society*, (131):2928–2934, 1984.

[144] Torrens. *Interatomic potentials*. Academic Press, New York, 1972.

Bibliography

[145] U. Messerschmidt. Introduction. In *Dislocation Dynamics During Plastic Deformation*, volume 129 of *Springer Series in Materials Science*, pages 3–9. Springer Berlin Heidelberg, 2010.

[146] V. Borovikov. *Multi-scale simulations of thin-film metal epitaxial growth*. PhD-Thesis, The University of Toledo, 2008.

[147] W. Eckstein, C. Garcia-Rosales, J. Roth, W. Ottenberger. Sputtering data. *Max-Planck-Institut für Plasmaphysik, Report IPP 9/82*, 1993.

[148] W. Sekkal, A. Zaoui, S. Schmauder. Nanoindentation study of the superlattice hardening effect at TiC(110)/NbC(110) interfaces. *Applied Physics Letters*, 86(16):163108, 2005.

[149] W.-T. Liu, D. R. McKenzie, W. D. McFal, Q.-C. Zhang. Effects of sputtering gas pressure on properties of silicon nitride films produced by helicon plasma sputtering. *Thin Solid Films*, 384(1):46–52, 2001.

[150] W. Xu, T. Fujimoto, I. Kojima. Preparation and characterisation of smooth and dense silicon nitride thin films. *Thin Solid Films*, (394):109–114, 2001.

[151] W.C. Oliver, G.M. Pharr. An improved technique for determining hardness and elastic modulus using load and displacement sensing indentation experiments. *Journal of Materials Research*, 7(6):1564–1583, 1992.

[152] W.C.D. Cheong, L.C. Zhang. Molecular dynamics simulation of phase transformations in silicon monocrystals due to nano-indentation. *Nanotechnology*, 11(3):173, 2000.

[153] Y. Chen, K. Matsumoto, Y. Nishio, T. Shirafuji, S. Nishino. Hetero-epitaxial growth of 3C-SiC using HDMS by atmospheric CVD. *Materials Science and Engeneering B*, (61-62):579–582, 1999.

Bibliography

[154] Y.H. Lin, P.F. Yang, S.R. Jian, Y.S. Lai. Molecular dynamics simulation of nanoindentation-induced mechanical deformation and phase transformations in monocrystalline silicon. *Nanoscale Research Letters*, 3:71–75, 2008.

[155] J. Ziegler. TRIM: The transport of ions in matter. Technical report, IBM Research, 28-0, Yorktown, New York, 10598, 1990.

Acknowledgement

At this point, I would like to thank all the people who participated in the making of this work.

- First of all, I thank Prof. Dr. rer. nat. Siegfried Schmauder for giving me the opportunity to be a respected member of his research group at the Institute for Materials Testing, Materials Science and Strength of Materials (IMWF), University of Stuttgart. I thank him both for his guidance as well as for the freedom he gave me to pursue my own ideas. Special thank goes to Prof. Schmauder for the thourogh revision of the manuscript and for being the main referee of my PhD work.

- I also thank Prof. Dr. rer. nat. Hans-Rainer Trebin for being a co-referee of my PhD work. In addition to his enrollment in my PhD work, I thank Prof. Trebin for managing the group at the Institute for Theoretical and Applied Physics (ITAP), University of Stuttgart, which has been continuousely developing the programe code (IMD) that made the research presented in this work possible.

- Priv.-Doz. Dr. Sven Ulrich has been supportive from the beginning on: I thank him for the close collaboration during our joint project, for his expertise and all the discussions during my PhD work which have sharpened my views. At last, I thank Sven for reading the manuscript and being a second co-referee in my PhD exam.

- Prof. Dr.-Ing. Michael Resch I thank for being the chairman of my PhD exam. A special thank goes to him also for managing the

Chapter 13. Acknowledgement

work group at the High Performance Computing Centre (HLRS), University of Stuttgart, his colleagues at the HLRS have always been supportive and competent in solving any technical issue that I had encountered during my work.

- Prof. Andreas Kronenburg, Prof. Peter Eberhard and Prof. Joachim Groß I thank for reading the dissertation manuscript and being a part of the board of examiners.

- Dr. Carlos Ziebert has become a good colleague and a valuable friend. Much of the work during our joint project was done in a close collaboration, long talks and discussions in person and over the phone. This would not have been possible without his friendly personality and scientific expertise. I thank him for the close collaboration, for reading the dissertation manuscript and providing valuable hints in improving the text.

- Dr. Jian Ye has been a silent but a hard working partner. He contributed in so many ways to the sucess of this work. I would like to take this opportunity to thank him for it.

- I have received much help from Dr. Peter Binkele: He has been very reliable and competent in reading the manuscript and making valuable propositions. I would also like to thank him for five years of very good collaboration.

- Dr. Christopher Kohler has put a great effort into making me an independant scientist and although he left the group in the early stage of my PhD work, he had open ears and a clear mind for many doubts I encountered later on. Thank you very much Christopher.

- Special thanks goes to Ms. Janine Lichtenberg, her results are partially published in this work. I would also like to thank Ellankavi Ramasamy, Julia Dürrwanger, Yousef Baroud, Wolfgang Verestek, Harsha Hossur Nagendra and Nooreldin Metwally Moussa. Thank you for your collaboration, your hard work and your motivation.

- I thank Kerstin Hilscher, Elke Kosthaus for their friendly personality and effort put in organizing the financial framework ai IMWF during the last years.

- Part of my time at the IMWF has been invested in IT administration within the work group. During this time, I had a pleasure to work with and learn from our colleagues from the MPA. I thank Ali Tabatabaei, Hans-Dieter Hilpert, Markus Münch, Simone Baisch, Christine Höfer and Siegried Winkler.

- At this point colleagues, who were not directly involved in my PhD work, but were important for providing a healthy and motivating working atmosphere need to be mentioned. I thank Ulrich Weber, Galina Lasko, Alejandro Mora, Haoyun Tu, Jing Wiedmaier, Andreas Reuschel, David Molnar, Marijo Mlikota, Martin Hummel, Steffen Hocker, Axel Krebs, Stefan Küster and Vinzenz Guski.

- Last but not least, I thank the German Science Foundation DFG for the financial support that made this work possible.

i want morebooks!

Buy your books fast and straightforward online - at one of world's fastest growing online book stores! Environmentally sound due to Print-on-Demand technologies.

Buy your books online at
www.get-morebooks.com

Kaufen Sie Ihre Bücher schnell und unkompliziert online – auf einer der am schnellsten wachsenden Buchhandelsplattformen weltweit! Dank Print-On-Demand umwelt- und ressourcenschonend produziert.

Bücher schneller online kaufen
www.morebooks.de

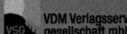

VDM Verlagsservicegesellschaft mbH
Heinrich-Böcking-Str. 6-8
D - 66121 Saarbrücken

Telefon: +49 681 3720 174
Telefax: +49 681 3720 1749

info@vdm-vsg.de
www.vdm-vsg.de

Printed by Books on Demand GmbH, Norderstedt / Germany